AREA GUIDELINES FOR SCHOOLS

Architects & Building Branch
Capital & Buildings Division
Schools Directorate
DfEE

London: HMSO

Acknowledgements

This publication has been prepared by the following team of DfEE Architects and Building (A&B) professionals, under the Chief Architect, Jeremy Wilson:

Executive editors: Beech Williamson and
                    Andy Thompson

Section specialists:

Section 2:   Beech Williamson
Section 3:   Robin Bishop
Section 4:   Lucy Watson
Section 5:   John Brooke

With contributions from Andrew Benson-Wilson, Tom Carden, Diane Holt, Jonathan Ibikunle, Sandra Legg, Helen Nichols and Alison Wadsworth.

The work was carried out in consultation with subject specialists from the Office for Standards in Education (OFSTED).

We are grateful to the many LEAs and individual schools who contributed to the background research, and to Learning Through Landscapes for their help with section 5.

Contents

Preface

This building bulletin gives non-statutory guidance on the provision of teaching and non-teaching accommodation for nursery, primary and secondary pupils. It also covers school grounds. Area Guidelines for Schools is published against the background of the removal of statutory teaching area and outdoor recreation area requirements for schools.

It is aimed at the early stages of school projects, when strategic decisions must be made about the buildings and site. Those responsible face a number of important choices in the way school premises are to be provided within available resources. This document is intended to help inform the way those choices are made.

There is an important balance to be struck between educational requirements, in terms of the curriculum and its delivery, the area of accommodation or land needed to support those requirements, and both the initial cost and running costs of that area. While the quality of school buildings and grounds and an imaginative plan for their development are clearly important to the school's function and its standing in the community, excessive area is to be avoided; it will not only cost more to provide but will represent an unnecessary drain on a school's budget year after year.

This building bulletin sets out a method for determining area needs and priorities when formulating a brief for the design of a school. It also provides the necessary points of reference for those most closely concerned with school building projects, whether new-build, extensions, adaptations or external works.

The advice given in this document is not prescriptive. Schools and local authorities need to formulate their policies in the light of their statutory duties and their own assessment of local resources.

Introduction

This building bulletin provides guidance on areas for primary and secondary school premises. It is aimed at anyone who is involved in the design of new schools or the remodelling of existing accommodation.

This document is relevant to premises in mainstream schools to which the Education (School Premises) Regulations 1996 apply[1]. It brings together area guidelines and general design advice for most types of school buildings and their grounds, and for all types of spaces, particularly teaching areas. This allows the needs of individual spaces to be seen in the context of the constraints of a reasonable gross area. Similarly, the overall building area can be seen in relation to the needs of the site, including the statutory requirements for playing field area.

This publication does not cover special schools, sixth form or tertiary colleges, or further education establishments. Advice on special schools and pupils with special educational needs in ordinary schools can be found in related building bulletins, as well as more detailed guidance on nursery provision, school grounds, and the accommodation needs of specific curriculum areas in secondary schools[2].

The design of any school building should also take account of some key considerations which are not covered in this publication, as they are dealt with elsewhere. These include:

- access for the physically disabled;
- environmental design;
- fire safety;
- constructional standards.

Primary and secondary area guidelines have been combined in one document. This has the advantage of demonstrating the common factors in the design process and also the common approaches to the use of space. Middle schools are not covered directly, but the relevant advice can be derived by combining the guidance for each age group involved.

It includes guidance on the areas of individual teaching spaces, the overall gross areas of buildings and site areas, generally in the form of graphs showing ranges of areas. These are based on simple formulae related to the age range and number of pupils accommodated. For each type of teaching space, an area is allowed for each pupil in the teaching group (G), depending on the activities and resources involved. This is added to a common area which is independent of the number of pupils, for instance to allow for circulation near the entrance. The gross area of buildings for a range of types of school is derived from similar formulae, using a common area of teaching and non-teaching accommodation plus an area per pupil related to the number on roll (N).

The advice given here is drawn from many years of research by Architects and Building (A&B) professionals and from their experience of designing and scrutinising school building projects.

This is an advisory document and the areas quoted and the methods used are designed to be flexible enough to cover most situations. The recommendations are reasonable within the context of finite capital resources. They are also intended as a point of reference in considering value for money in all school projects.

Local authorities and individual schools will, of course, establish their own building priorities in the light of the funds they have available for capital work and any conditions which may be attached to these funds.

Relationship to Capacity Assessment

This publication takes account of the recognised methods of calculating the capacity of either a primary or secondary school. Annexe D to Circular 6/91 applies to primary and annexe A to Circular 11/88 applies to secondary. These are commonly known as the 'MOE (More Open Enrolment) formulae' for calculating capacity from a schedule of existing or proposed accommodation. Both remain in force, so when devising the schedule of teaching accommodation for a particular size of school from this document, it is useful to check on the capacity of such a schedule, calculated using the relevant MOE formula.

Notes
1: The Education (School Premises) Regulations 1996 came into effect from 1 September 1996. They are referred to throughout this document as the School Premises Regulations.
2: Recent A&B publications are listed in the bibliography. For guidance on special schools, reference should be made to Building Bulletin 77: Designing for pupils with special educational needs: Special schools, HMSO 1992.

Using This Document

The guidance in this document follows the steps that designers and school planners can take to identify the appropriate areas for all mainstream schools.

Section 1 identifies the approximate overall area for the school buildings. Because capital costs for school buildings are directly related to area, and the recurrent costs for every square metre of accommodation remain a drain on schools' budgets, overall area must be constrained.

Section 2 can be used to establish the number and type of teaching spaces needed to support particular curriculum or staffing models.

Sections 3 and 4 provide more detailed information on the individual spaces required.

Section 5 deals with the site area and layout. Reference to this section will be useful at different stages in the design process. It should help in choosing a site, in locating a new building or extension, and in planning a layout of the main external features.

Appendices 1 to 3 show worked examples of schedules of accommodation for 5-11 primary, 11-16 and 11-18 secondary schools, together with a discussion of the options for each school. Appendix 4 summarises the area formulae for various types of school.

Stage 1: Identifying the Boundaries of what is Possible or Affordable

Section 1 gives advice on a range of figures for the total gross area of buildings for schools of different types and sizes. These figures have been derived from observation of good practice. This section also gives advice on the proportions of the overall area that should be given to teaching[1] and non-teaching accommodation.

The range of figures quoted provide for new schools and, whilst allowing for local circumstances and priorities, are likely to represent value for money. The figures can also be used as a reference for existing schools when remodelling or extension is being considered.

Stage 2: Establishing Priorities

Section 2[2] explains a method of calculating the quantity and type of timetabled teaching accommodation required through an analysis of the existing or proposed curriculum and its delivery. Although this method was initially developed for secondary schools, it can also be used for other types of schools, particularly the larger ones. Appendices 1 to 3 give detailed examples for different types of school.

The methods proposed require a close collaboration between client and designer in identifying user needs, assessing their implications and determining priorities.

The essential elements are to establish the distribution of pupils' and teachers' time and then to calculate the numbers of each type of space required, recognising the need for some flexibility in the use of space.

Stage 3: Deciding on Space Sizes

Sections 3 and 4 give detailed information for the next part of the process; calculating the sizes of individual spaces. A range of areas of teaching space are illustrated for different group sizes and activities. The characteristics of spaces at the top and bottom of the ranges are discussed.

Stage 4: Working within Constraints

The process of working through these steps, from broad constraints to detailed design considerations, provides a framework for the designer. However, it is very unlikely that there will be a perfect fit between the overall area, the demands of the timetable and the areas required for different subjects. It may be that after identifying spaces from an analysis of the curriculum in section 2, combined with desirable results from sections 3 or 4, the constraints of section 1 are breached. Compromises will then have to be made by identifying priorities for the individual school.

Notes
1: For nursery provision, the teaching area includes playrooms.
2: This section develops the methods first described in Design Note 34, Area Guidelines for Secondary Schools 1983.

Section 1: Gross Area of Buildings

This section outlines a range of values for the total gross area of buildings which constitute value for money in new schools. These area guidelines are followed throughout this document.

1.1 The total gross area[1] of the buildings of a school is a significant figure. Capital costs are directly related to the provision of floor area in that it costs more to build a larger building. For every square metre of area in excess of real need there will be the capital cost of providing it. Added to this will be a recurrent figure, equivalent to between 4% and 6% of the capital cost, to be paid every year for cleaning, maintenance, heating, lighting, insurance and rates.

1.2 Generally the gross area per pupil will be higher for a small school than for a large one, because of economies of scale (primary schools may need a hall, head's room, medical room, regardless of the number of pupils). The area per pupil of secondary schools will be higher than primary schools due to their more complex organisational structure and their need for a wide range of specialist spaces to support the delivery of the curriculum at that level. Post-16 pupils may require more area for similar reasons.

1.3 The ranges of figures given in this section for total gross area represent a realistic balance between what is desirable and what is likely to be possible within available resources. Whilst they are intended to apply to new schools, they can also serve as useful yardsticks for existing schools. When extension or adaptation projects are considered, it is important that the overall area of the school does not increase above what is necessary. The closeness of the 'fit' that can be expected depends on both the suitability of the existing accommodation and the extent of the adaptation work proposed.

1.4 A gross area target can be determined from the range by assessing the:

- current and projected numbers on roll;
- type of school;
- existing site and accommodation;
- resources available.

1.5 The next step is to target the proportion to be assigned to teaching area (this is a significant factor affecting decisions outlined in section 2). It is important that teaching area forms as large a proportion of the total area as possible, whilst at the same time allowing enough space for non-teaching accommodation, in particular that required by the School Premises Regulations and health and safety legislation[2]. As a general rule, schools should be aiming roughly at a 60:40 split between teaching and non-teaching area. In nursery schools this is likely to be nearer a 50:50 split.

Gross Area Formulae

1.6 The following guidance for total gross areas has been derived from observation of primary and secondary schools which are delivering a full curriculum within appropriate accommodation. The areas for a range of common school sizes are within defined formulae. The gross areas of primary and 11-16 secondary schools are illustrated graphically.

1.7 It is worth noting that the circumstances of each school are different and there will be some cases where the area will be outside the range outlined here, particularly if there is significant provision for special educational needs. The position selected within the graphs depends on available capital resources and local circumstances such as curriculum priorities and staffing limits. However, the whole range is considered to represent reasonable value for money.

Nursery Provision

1.8 Nursery pupils may be in nursery classes or units attached to primary schools, or in separate nursery schools. Nursery classes and units may use some of the non-teaching facilities of the main school. The accommodation needs of nursery pupils, which are different from those of primary pupils, are summarised in paragraphs 3.57 to 3.76, although at the time of writing research into nursery accommodation is continuing.

Notes
1: The gross area, defined in the glossary, is the total floor area of all the school's buildings, measured to the inside face of the external walls, but including the area of internal walls.
2: Relevant health and safety publications are listed in the bibliography.

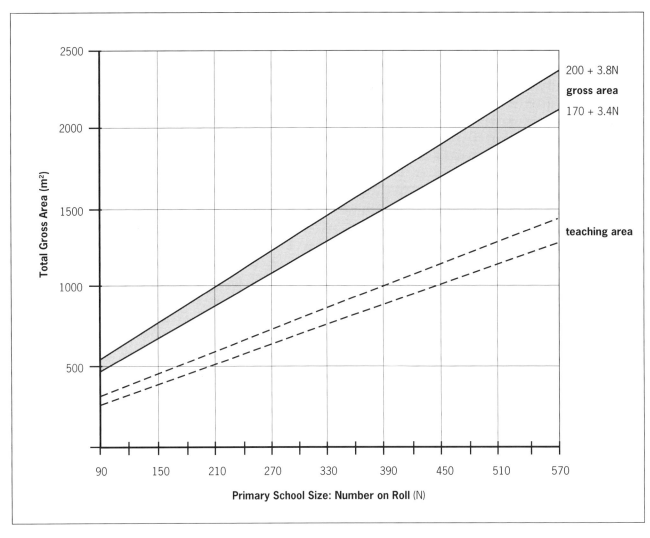

Figure 1.1: primary schools: graph showing gross area guidelines representative of value for money in new 5-11 primary schools. The broken lines indicate a range of likely teaching area.

1.9 Because of the small scale of nursery provision and the possible variation in its relationship, if any, with an associated primary school, it is not appropriate to give ranges of areas for general use. However, a reasonable gross area target to accommodate a 26 place nursery attached to a primary school would be around 100m² (100 square metres).

Primary Schools

1.10 Recent observation has shown that there is a case for using the same area standards for infant and junior pupils[1]. For the purpose of this document the gross area standards are the same for all primary age groups (referred to as Reception, Key Stage 1 (KS1) and Key Stage 2 (KS2) in the National Curriculum).

1.11 Figure 1.1 shows a range of total gross areas for any type of primary school of 90 to 570 pupils as defined by the formulae attached (where N is the total number on roll).

1.12 For schools with more than 570 pupils some continuing economies of scale might be expected, but there is insufficient evidence to support a precise area recommendation. With schools for fewer than 90 pupils, area standards will depend very much on individual decisions concerning the provision of a hall and other shared accommodation. Once again it is not possible to make general recommendations of area in these circumstances.

1.13 The 60:40 rule of thumb for the proportions of teaching and non-teaching area has been tested against new primary schools built in recent years and through studies of area schedules for existing schools. The results show that primary schools with 240 or more pupils can achieve a teaching area of at least 60% of

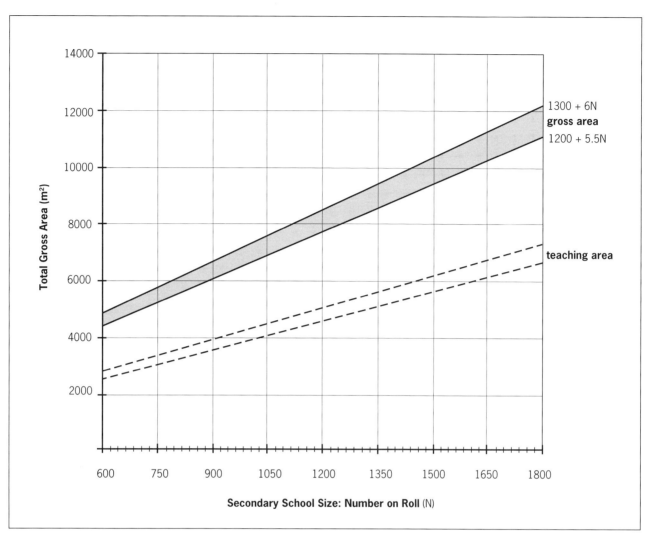

Secondary School Size: Number on Roll (N)

gross. The broken lines in figure 1.1 indicate a range of teaching areas based on this approach. A significantly higher figure is possible for much larger primary schools. In the smaller schools this percentage can fall to around 57% (or even less in very small schools). However, it is generally only by achieving over 60% of the gross that it becomes possible to secure significant improvements to teaching area.

Secondary Schools

1.14 The range of total gross areas for 11-16 schools above 600 pupils is shown in figure 1.2, together with the determining formulae (where N is the total number on roll). Individual judgements must be made about any diseconomies of scale in schools for less than 600 pupils.

1.15 As in primary schools, the percentage of gross area that should be targeted for teaching area is affected by the school size. For schools with 1350 pupils the target should be 60% or more, whilst for schools with 750 pupils 58% to 59% would be nearer the mark. The broken lines in figure 1.2 indicate a range of teaching areas based on this approach.

Sixth Forms

1.16 Sixth form pupils require a larger area, so the following formulae can be added to the formulae in figure 1.2, where n is the total number of pupils in the sixth form and N is the total number on roll, *including the sixth form*:

upper line   3n

lower line   2.5n

1.17 The formulae for the total gross area for an 11-18 school, applied to all pupils, therefore become:

upper line   1300 + 6N + 3n

lower line   1200 + 5.5N + 2.5n

Figure 1.2: secondary schools: graph showing gross area guidelines for 11-16 schools representative of value for money in new schools. The broken lines indicate a range of likely teaching areas.

Note
1: Although younger, smaller children might be expected to take up less space than older ones, they may need as much due to more extensive activities and larger equipment. For example, younger children engaged in practical activities tend to occupy more area, in part due to a less developed ability to work in restricted areas. Also, the amount of movement of young children is the same or higher than that of their elders. Resources also take up much the same space. The juniors may have more specialised resources and be physically larger, but the infants tend to use larger scale equipment.

1.18 This area is based on smaller teaching spaces being provided for small sixth form groups when the frequency of use justifies exclusive use.

Other Types of School

1.19 In the secondary sector, different area standards are needed for the different key stages (KS3 and KS4), which are in turn higher than the area demands of the primary sector (Reception, KS1 and KS2). Increasing differentiation of the curriculum usually means that teaching groups are smaller for older pupils. Thus, KS3 pupils use the same teaching space more economically and require a lower gross area than the older pupils. This is reflected in pupil:teacher ratios commonly varying between the key stage groups, with a reduction at KS4. Similarly, post-16 pupils require more area than KS4 pupils.

1.20 Although middle schools may be 'deemed primary' or 'deemed secondary', both have pupils of primary and secondary age and need to accommodate work at both KS2 and KS3. In gross area terms the overall constraints are determined by a proportion that reflects the allowance for the number of pupils in each key stage.

1.21 A range of possible gross areas of middle schools, 13-18 secondary schools and schools covering other age ranges can be estimated by using the following formulae for each KS in proportion to the number of pupils in each sector in the school. The additional formulae for sixth form (paragraph 1.16) can be added if the school has a sixth form.

KS1 and KS2 total gross area (m²)

| upper line | 200 + 3.8N |
| lower line | 170 + 3.4N |

KS3 total gross area (m²)

| upper line | 700 + 5.7N |
| lower line | 600 + 5.2N |

KS4 total gross area (m²)

| upper line | 2200 + 6.45N |
| lower line | 2100 + 5.95N |

1.22 In all cases, N is the total number on roll, including all reception pupils, KS1 through to KS4 classes and any sixth form. The formulae do not apply to nursery schools, sixth form colleges or special schools.

Gross Area per Pupil

1.23 The range of gross area per pupil possible for a school can be derived by dividing the total gross area by the total number on roll (N). In practice this can be expressed with formulae as simple as those above. For instance, the lower limit of the gross area per pupil for an 11-16 school is:

$$\frac{(1200 + 5.5N)}{N}$$

$$= \frac{1200}{N} + 5.5$$

Gross area per pupil formulae can be used for other age ranges in the same way.

1.24 Appendix 4 shows formulae for total gross area ranges and for the gross area per pupil for a variety of types of school, based on this system.

Section 2: Deriving an Accommodation Schedule

This section describes the method for deriving a schedule of accommodation. The section uses an 11-16 secondary school as an example, explained in more detail in appendix 2, but the same principles are used in the primary and 11-18 examples in appendices 1 and 3. The overall schedule comprises three types of space; **timetabled teaching area, non-timetabled teaching area** and **non-teaching area.** The number of timetabled spaces can be derived using a **curriculum analysis**, to calculate the demand for spaces based on the school's curriculum.

2.1 The simple calculations described under 'Curriculum Analysis' can be used to identify the spaces needed for separate subjects. More usefully, the overall analysis summarised in 'Creating a Schedule for the Whole School' can be used to show the future demand for spaces based on the likely future curriculum model and number on roll.

2.2 The working method and arithmetic involved are straightforward and can be used to demonstrate the effects of curriculum balance, level of staffing, teaching group size and the nature of the subjects covered on the number of spaces required. For primary schools, it may be useful to use a curriculum analysis to consider those spaces that are to be shared by all classes, such as the hall, and therefore need to be timetabled. In secondary schools the organisation is usually more complex and most spaces are timetabled.

2.3 The majority of teaching spaces in secondary schools are timetabled. This section deals firstly with the principles of curriculum analysis, which can be used to identify the number of timetabled spaces required. The other areas in the school can then be added, resulting in the overall schedule of accommodation.

Curriculum Analysis

2.4 Curriculum analysis is particularly useful in secondary schools, but it can also be applied to middle and primary schools (see paragraphs 2.42 and 2.44). The method is flexible enough to create an area schedule for new school buildings, and also to assess the need for spaces as the curriculum or numbers on roll change. In an existing school it can highlight the areas where refurbishment or a change of use will improve the delivery of the curriculum.

General Principles

2.5 At the simplest level, the number of teaching spaces required in any school will relate to the number of teachers and the proportion of their time that they spend teaching.

2.6 The amount of time that teachers spend in face to face teaching will be a proportion of their working time at school, identified as the *contact ratio*. That is the average proportion of a full timetable week that teachers spend teaching a timetabled group. For example 0.78.

2.7 The calculation of the number of teaching spaces required must therefore include the contact ratio to identify the time when teaching is actually taking place. For example, in the 900 place 11-16 secondary school used here (and in appendix 2), the full-time equivalent (FTE) number of 47.6 teachers are employed at an average contact ratio of 0.78. The average number of teachers teaching at any one time would be:

$47.6 \times 0.78 = 37.13$

2.8 If all timetabled teaching rooms were used for 100% of the time, this number would be the number of rooms required. However, it would be organisationally impossible to use all spaces for all of the time, so a reasonable *frequency of use* should be identified. A more realistic total can be calculated if an average frequency of use of about 85% is assumed:

$37.13 \div 0.85 = 43$ spaces approximately.

2.9 This allows for some general spaces to have a higher frequency of use (around 90%) and the use of specialist rooms to be more variable (often around 75% to 85% of the time). It also assumes that some PE will take place outdoors. This is a total of timetabled spaces and does not include ancillary teaching spaces such as the library.

2.10 As this example illustrates, the number of timetabled spaces is usually fewer than the number of teachers, so most spaces will not be used solely by one teacher.

2.11 Although the above example is a useful overall check, it is important to know what types of space are needed and how many there are of each.

Types of Timetabled Space

2.12 Timetabled teaching spaces in any school include a variety of types to suit different activities. In secondary schools the types of spaces needed usually relate to the subjects taught. For instance, science is usually taught in a science laboratory.

2.13 Some subjects, such as design and technology, require a variety of specialist spaces to suit the different activities, such as in food or multi-materials. Other subjects require general teaching classrooms which need not be subject specific.

Proportion of Time

2.14 If it is assumed that the different types of space required relate directly to the subjects taught, then the demand for space can be identified by the time that a subject is taught. For instance, if all science (and only science) is to be taught in science laboratories, the number of laboratories can be calculated by identifying the proportion of teacher time spent teaching science.

2.15 It is also useful to identify the demand for rooms across subject boundaries, such as the need for an IT room which may be timetabled for use by a number of subjects. This is particularly relevant to sixth-form provision and primary schools, and is covered in the examples in appendices 1 and 3.

The Timetable

2.16 Secondary school teaching is predominantly organised in a timetable, in which a week or sometimes a fortnight is divided into a number of *periods*. The week could be divided into any number of periods, but so long as the same unit is then used in all calculations this will allow a convenient division of taught time.

Determining the Number and Types of Rooms

2.17 An analysis of the timetable can determine three key elements of information:

- a breakdown of the total number of *teacher periods* (i.e. one teacher teaching a lesson for one period);

- the number of *pupil periods* spent in each subject (i.e. one pupil being taught for one period);

- the *average group size* of each subject (i.e. the average size of a class group being taught at any one time).

Using Teacher Periods

2.18 As discussed above, the number of spaces needed relates to the number of teachers teaching at any one time. This can be measured in the form of teacher periods. The total number of teacher periods in any subject divided by the number of periods available in the week (or fortnight) is the average number of lessons being taught in that subject at any time, and therefore the number of spaces required if used 100% of the time.

2.19 For instance, if 233 teacher periods are spent teaching science in a timetable of 40 periods per week, the calculated number of spaces for science would be:

no. of teacher periods ÷ periods per week

$$= 233 \div 40 = 5.83$$

This calculated number can then be rounded up and the frequency of use can be found by dividing the calculated figure by the rounded figure. Therefore 5.83 could be rounded up to:

6 spaces at 97% (5.83 ÷ 6) frequency of use, or

7 spaces at 83% (5.83 ÷ 7) frequency of use.

Using Pupil Periods and Average Group Sizes

2.20 The total amount of time that pupils spend on any subject can be measured using pupil periods. The number of teacher periods multiplied by the average group size equals the number of pupil periods.

teacher periods x group size = pupil periods.

2.21 In this case the average group size would be estimated bearing in mind the range of group sizes that may be taught. The average group size may be difficult to determine precisely, so it may need to be reconfigured in the final calculation. The maximum group size can help to determine the ideal size of the room (see paragraph 2.35).

2.22 It is often useful to use pupil periods to calculate space numbers, when only the percentage of curriculum time to be aimed at and the approximate group sizes are known. The curriculum percentages are a proportion of pupil time and not the proportion of teacher time unless the group size remains the same for all subjects and age groups.

2.23 For example, if 7.5% of the total timetabled curriculum time is spent on design and technology, the number of pupil periods spent will be 7.5% of the total (i.e: the number on roll x the periods per week), or:

7.5% x (900 x 40) = 2700

This number of pupil periods can be divided by an approximate average group size, in this case 20, to give the teacher periods. The result can be further divided by the periods per week to calculate the number of spaces, as in paragraph 2.19. This would be:

$$7.5\% \times \frac{900 \times 40}{20 \times 40} = 3.38$$

or 4 spaces at 84% frequency of use

or 5 spaces at 68% frequency of use.

A Short Cut

2.24 As the periods per week figure occurs on the upper and lower lines of the calculation, it can be cancelled out, leaving a simple formula for calculating the number of spaces required for any subject on the basis of curriculum percentage:

$$\frac{\text{curriculum percentage x N}[1]}{\text{average group size}} = \text{no. of spaces}$$

The same example would be:

$$7.5\% \times \frac{900}{20} = 3.38$$

2.25 One advantage of using this method is that the number of spaces can be estimated using the minimum of information. Also, the demand for an increasing number on roll, with a similar average group size and the same percentage of curriculum taught, can easily be determined by using the same calculation with a revised number on roll. So if the roll of the previous example was to increase to 1200, the number of spaces demanded would be:

$$7.5\% \times \frac{1200}{20} = 4.5$$

or 5 spaces at 90% frequency of use

or 9 spaces at 75% frequency of use[2].

2.26 If this method is used for a whole school (as in the primary example in appendix 1) it should be checked that the total calculated number of spaces is not less than the number of teachers multiplied by the contact ratio, to ensure that there are sufficient teachers to teach in the group sizes chosen.

Creating a Schedule for the Whole School

2.27 If the calculations above are done in a spread-sheet format for the whole timetable, the total number of the different types of spaces required can be calculated. It is usually important to be aware of the whole picture as changing one area of the school will often have implications elsewhere. For example, if the percentage of design and technology is increased and the number on roll and total teaching time remain the same, then other subjects must be reduced as a proportion of the curriculum. It is important to identify this space

Notes
1: N = number on roll.
2: In practice, this may be an estimated figure, as the actual average group size may vary as the year group size increases (eg 180/9 = 20 but 190/10 = 19). However, both would be rounded up to the same number of rooms.

Section 2: Deriving an Accommodation Schedule

Figure 2.1: example of a schedule of accommodation for a 900 place 11-16 school, including a summary of curriculum analysis (steps 1 to 4), non-timetabled and non-teaching area (step 5). This is an alternative solution to the curriculum analysis detailed in appendix 2.

a = periods per week	40	
b = total number on roll (NOR)	900	
c = FTE number of teachers	47.6	
d = contact ratio	0.78	
e = total pupil periods (a x b)	36000	= total v
f = total teacher periods (a x c x d)	1485	= total w

lower line of gross area = 6150m²
upper line of gross area = 6700m²

	STEP 1				STEP 2	STEP 3		STEP 4		
subject	percentage of curriculum u	total pupil periods v	total teaching periods w	average group size x	number of spaces calculated y	number of spaces adjusted z	frequency of use	maximum group size	average area per space m²	total teaching area m²
method explanation	v as percentage of total	from curriculum breakdown	from curriculum breakdown	v / w	w / a	based on reasonable freq. of use	y as prcentage of z	based on curriculum breakdown	from section 4 guidelines	z x average area
English	12.5%	4500	185	24.3	4.63					
mathematics	12.5%	4500	185	24.3	4.63					
modern foreign languages	11.5%	4140	169	24.5	4.23					
humanities	10.0%	3600	148	24.3	3.70					
religious education	5.0%	1800	74	24.3	1.85					
PSE	1.5%	540	21	25.7	0.53					
general studies	2.0%	720	32	22.5	0.80					
TOTAL GENERAL	**55.0%**	**19800**	**814**	-	**20.35**					
standard @ 77.5%		15345	631	24.3	15.77	18)	88.5%	30	50	**900**
large @ 22.5%		4455	183	24.3	4.58	5)		30	62	**310**
IT/business studies	**2.5%**	**900**	**45**	**20.0**	**1.13**	2	56.3%	25	72	**144**
science	**15.5%**	**5580**	**233**	**23.9**	**5.83**	7	83.2%	30	85	**595**
design and technology	**7.5%**	**2700**	**135**	-	**3.38**					
food @ 21.5%		581	29	20.0	0.73	1	72.6%	21	103	**103**
m-m/graphics @ 35.5%		959	48	20.0	1.20	2	59.9%	21	103	**206**
PECT @ 21.5%		581	29	20.0	0.73	1	72.6%	21	88	**88**
textiles @ 21.5%		581	29	20.0	0.73	1	72.6%	21	84	**84**
art	**4.0%**	**1440**	**52**	-	**1.30**					
2D art @ 60.0%		720	26	27.7	0.78	1	78.0%	30	91	**91**
3D art/textiles @ 40.0%		720	26	27.7	0.52	1	52.0%	30	109	**109**
music	**4.0%**	**1440**	**52**	**27.7**	**1.30**	1.5	86.7%	30	70	**105**
drama	**2.5%**	**900**	**34**	**26.5**	**0.85**	0.5	90.0%	30	91	**46**
PE (indoor)	**4.5%**	**1620**	**60**	**27.0**	**1.50**	2	75.0%	30	260	**520**
games (outdoor)	**4.5%**	**1620**	**60**	**27.0**	**1.50**	(external space)				
TOTAL	**100%**	**36000**	**1485**		**37.13**	**43**				**3301**

STEP 5					
NON-TIMETABLED SPACES	special educational needs		1	6	21	**21**
	library resource centre		1	-	143	**143**
	local/IT resource areas		2	15	28	**56**
	careers area		1	5	13	**13**
	FLA/seminar space		1	3	7	**7**
	music group rooms		5	4	8	**40**
	darkroom		1	3	10	**10**
	heat treatment bay		1	4	15	**15**
	kiln		1		4	**4**
	assembly hall (used for drama)		1	40.0%	-	260	**260**
TOTAL TEACHING AREA						**3870**
NON-TEACHING AREA	staff accommodation	@ 5.5% of gross				338
	pupils' storage/washrooms	@ 5.0% of gross				308
	teaching storage	@ 5.0% of gross				308
	catering facilities	@ 4.5% of gross				277
	ancillary/circulation/partitions	@ 20.0% of gross				1230
TOTAL GROSS AREA						**6330**

as it may be enough to allow the technology suite to be enlarged without requiring new building.

2.28 The curriculum analysis of the whole school can be added to the untimetabled teaching area and non-teaching area to give a schedule of accommodation for the school. In an existing school, this can be compared to the actual number, size and type of spaces to point to the areas that may require alterations to suit a changing curriculum or number on roll.

A Schedule Example

2.29 The three schedule examples in the appendices each follow five steps. The five steps are briefly outlined here, using the simple 900 place, 11-16 example in appendix 2, summarised in figure 2.1, in which the total pupil periods and teacher periods and the average group size for each subject are known.

Step 1: Distribution of Time

2.30 The first step is to identify, for each subject, the total pupil time (in pupil periods: column v), the number of times each group must meet (in teacher periods: column w) and the average group size (column x).

$$\frac{\text{Pupil periods}}{\text{teacher periods}} = \text{group size}$$

If two values are known, the third can be calculated (normally the group size).

2.31 The total pupil periods should equal the periods per week (a) x the total number on roll (b). Similarly, the total teacher periods should equal the periods per week (a) x the FTE number of teachers (c) x the contact ratio (d).

Step 2: Calculated Number of Spaces

2.32 Using the information in step 1 and the formulae outlined earlier (see paragraph 2.19) of

no. of teacher periods ÷ periods per week

the number of spaces demanded can be calculated (column y).

Step 3: Rounded Number of Spaces

2.33 At this point, the actual number and type of spaces required can be decided, based on a reasonable frequency of use. If this frequency is too little, a space may need to be used for more than one activity; if too much, more spaces may be required. The maximum and minimum frequencies of use will depend on the school's preferences within each subject.

2.34 In this example (figure 2.1), two IT spaces are required but the frequency of use is only 56%. The school has accepted this as these rooms will also be used as untimetabled, bookable resource areas for whole classes or smaller groups. The initial rounded number of total general teaching spaces was 21, with a high frequency of use of 97%, so a further two classrooms were added to allow an average usage of 88.5%. PE demands two spaces at 75% use, so a sports hall is to be used. Music requires two rooms at 65% usage, but by sharing one as a drama studio and by using the assembly hall for other drama activities for up to 40% of the time, only two spaces are required for both music and drama, used for 87% to 90% of the time. Other options are available and are discussed in appendix 2.

Step 4: Space Sizes

2.35 The area of each space must be decided on the basis of the largest group size that will use it and the activities that will take place within it, using section 4.

Step 5: Additional Areas

2.36 Untimetabled teaching areas such as the library may be partially timetabled but will mainly provide an area that is available for pupils to 'drop-in' and access shared resources. Other untimetabled spaces are listed in section 4. All such areas should be added to the list of teaching spaces, to give a final total teaching area. A full schedule of accommodation for a school will also include non-teaching area, as illustrated in figure 2.1.

Overall Adjustments

2.37 Once the initial list of spaces is complete, some quick checks can be applied:

- the total teaching area should be a reasonable proportion of the gross area (around 60%) as outlined in section 1;

- the average frequency of use of the final number of timetabled teaching spaces should be between 84% and 89%[1];

- the calculated average group sizes should fall comfortably within the chosen maximum group sizes for rooms;

- the capacity of the final schedule, calculated from the 'MOE formula', should be the same as or a little more than the planned number on roll, to allow for some flexibility in admissions;

- there should be an adequate number of spaces suitable for registration.

Other Types of School

2.38 Primary and 11-18 school examples are discussed in appendices 1 and 3. The principles are the same, the notable difference being that spaces are less likely to be specific to one subject, but more related to a range of activities that may be carried out in a number of subjects. Similarly, any type of school serving any age range can be analysed using curriculum analysis.

2.39 Many of the complications envisaged in such schools can be simplified by calculating the number of spaces at step 2, using different methods if convenient, for parts of the school, then adding them together before step 3. For instance, in appendix 1 a primary school has an infant and junior department with different lengths of school day. Appendix 3 details a secondary school with a sixth form with a complex curriculum breakdown. In both cases, the calculated number of spaces (step 2) has been derived for the different age groups and then added together. Other types of school with similar situations, such as middle schools, can also use this system effectively.

11-18 Schools

2.40 Many sixth form courses such as GNVQs (General National Vocational Qualifications) demand a variety of activities. These may be best accommodated in existing specialist rooms, shared with other courses or 11-16 subjects. The curriculum breakdown in appendix 3 totals the demand for each type of space in all subjects or courses.

2.41 In sixth forms, group sizes will often be smaller and the ability to identify small spaces for solely sixth form use can be useful.

Primary Schools

2.42 Teaching areas that are open and used by only a few classes may be shared effectively on an ad-hoc basis and can be included in the overall basic teaching area. Teaching spaces which are shared by a number of classes, such as the hall, studio, small group/withdrawal spaces and discrete practical spaces (identified as timetabled supplementary spaces in section 3), may need to be timetabled or booked. The curriculum analysis method can be used to quantify the demand for such spaces.

2.43 Small primary schools will usually have a small and predictable number of timetabled spaces. Larger primary schools or, as in appendix 1, those increasing in size, will find a curriculum analysis more useful.

Middle Schools

2.44 The middle school is a hybrid between the primary and secondary curriculum analysis. As discussed in paragraph 2.39, a primary and secondary analysis can be done separately up to step 2 of the process and then brought together.

Note

1: If a common contact ratio is known, a general check of the number of timetabled spaces to the number of teachers can be useful.

The final number of spaces = the FTE number of teachers x the contact ratio ÷ average frequency of use.

So, for instance, if the contact ratio is expected to be about 0.78, the number of timetabled spaces should be between 87.5% and 92.5% of the FTE number of teachers (0.78 ÷ 89% and 0.78 ÷ 84%).

Section 3: Primary Accommodation Guidelines

This section provides guidance on teaching and non-teaching areas in primary schools and on nursery accommodation. For primary schools, the teaching accommodation can be considered under three categories. The **basic teaching area** includes individual class bases and any shared teaching areas, for instance for practical activities. **Timetabled supplementary areas** include the hall and, if provided, discrete specialist practical areas, group rooms and studios. **Other supplementary teaching areas** include the library and resource areas. **Non-teaching areas** and **nursery provision** are covered at the end of this section.

3.1 In practice, there may be some overlap of the categories above, especially in smaller primary schools. Features of any well designed primary teaching area will include:

- good circulation (bearing in mind the active nature of many primary tasks, and the presence of non-teaching assistants or other adults);

- a configuration of furniture that assists supervision;

- good sightlines (especially between teacher, pupils and board);

- flexible furniture arrangements, with mobile items where possible;

- display surfaces (horizontal and vertical) for pupils' work, natural objects and both small and large artefacts;

- lighting, heating, water and power points positioned for convenience and safety;

- resources displayed for ease of pupil access.

3.2 Most resources stored in teaching areas are for the pupils to select and use and should therefore be at low level. Only valuable or hazardous items need be kept in locked cupboards or at high level. Full height storage, whether in the form of furniture units or walk-in stores, is classified as *teaching storage* and counts as non-teaching area for the purpose of area calculations.

Teaching Area

3.3 An area of 1.8m² per pupil, not including the hall and library, can support a range of teaching activities, but some compromises may be necessary. Layouts may be crowded and any practical work would need careful management. At 2.1m² per pupil, activities can be accommodated more comfortably, with less time spent rearranging furniture or getting work out and putting it away. Examples of spaces designed to these different levels are discussed in paragraphs 3.12 to 3.16.

Basic Teaching Area

3.4 Basic teaching area includes class bases and any shared teaching areas, other than supplementary teaching areas.

Class Bases

3.5 A base is required for every class or group of children within the school. Whether in the form of an enclosed classroom, or as part of a more open plan area, it should be large enough to gather at least 30 pupils together for registration, listening, discussion and for whole class teaching. An area of at least 35m² is likely to be needed for:

- adequate table space for every pupil, whether arranged in groups or rows;

- the teacher's workstation (not necessarily a desk).

3.6 Other basic ingredients for class teaching include:

- free floor space, for gathering pupils together and for space-consuming work on the floor;

- a book corner with room to browse;

- one computer workstation for Key Stage 1 (KS1) classes, and two for KS2 (suitably sized and sited)[1];

- facilities for practical work (see paragraph 3.9).

These activities need not be accommodated in the class base itself.

Note

1: The video and leaflet 'Making IT Fit' (DfEE 1995) provide guidance on accommodating computers in schools.

13

Shared Teaching Area

3.7 Many teaching activities can also be carried out in shared areas. If designed carefully, these areas can accommodate different approaches to teaching and foster collaboration. There is less duplication of resources and a more efficient use of space may be possible.

3.8 The configuration and environmental conditions are critical for shared areas. If badly designed, they can be noisy and distracting. A shared area should not be treated simply as a generous circulation route, but should offer opportunities for a reasonable variety of activities and be within sight and reach of the class teacher.

Practical Activity

3.9 Some subjects contain a practical element, such as maths, science, technology and art. These activities may take place in the basic teaching area or in a specialist area. In addition to the ingredients listed above, the following may be required:

- an area for wet or messy activities, with a deep sink and suitable floor finish;

- a dry practical area for work such as making and testing;

- sufficient power points;

- appropriate furniture, such as fixed work tops, workbenches and water or sand trays.

Access to a range of practical resources is also essential.

3.10 The extent of the area for practical activity depends upon a number of factors, including the level of adult supervision and the size of the group doing practical activity. These apply whether the practical areas are provided in enclosed classrooms, areas shared by two or more class bases, or a specialist practical area shared by all classes in a key stage or the whole school (as described in paragraph 3.18). These organisational considerations have to be weighed against the physical and gross area constraints. The chief advantage of practical activity in the basic teaching area

is that space is available for other activities at other times.

3.11 Figure 3.4 illustrates a range of recommended areas for practical activity, whether in the basic teaching area or in a specialist practical area.

Ways of Organising Basic Teaching Area

3.12 The design of basic teaching accommodation can vary from a collection of entirely self-contained classrooms to a more open plan system with a high proportion of shared areas and small class bases. Each approach can work well as long as they meet the school's current needs and are flexible enough to allow for changes in the future.

Enclosed Room

3.13 Self-contained classrooms for 30 pupils offering a reasonable range of teaching activity would be in the range of 54m^2 to 63m^2. Examples of these are illustrated in figure 3.1. An enclosed room allows the teacher to monitor pupils more closely and provides more autonomy and privacy (with advantages for music or noisy activities). However, there is less opportunity for co-operation and shared supervision.

Open Plan

3.14 At the opposite extreme, an open plan base will rely on screens and furniture to create any form of enclosure for each class. This is perhaps the most difficult organisation of space to design and manage effectively. The benefits of flexibility through co-operation and shared supervision tend to be offset by noise and visual disturbance.

Semi-open Class Bases

3.15 Bases that open onto a shared area can support a variety of teaching methods, though it should be possible for teachers to screen class bases off when more privacy is required. Such arrangements can also give easier access to shared

Enclosed Classroom Examples (figure 3.1):
Two sizes of enclosed classroom are illustrated below. The smaller room is 54m² (1.8m² per pupil) and can support a range of activities although some compromises have been made. For instance, there is only one computer workstation. However, this may be compensated for by shared basic teaching area or supplementary spaces elsewhere in the school. In the larger room of 63m² (2.1m² per pupil) activities can be accommodated more comfortably and there is also room for a book corner. However, providing such an area for all classes would limit the area available for supplementary spaces, so the classroom would need to accommodate most practical activities.

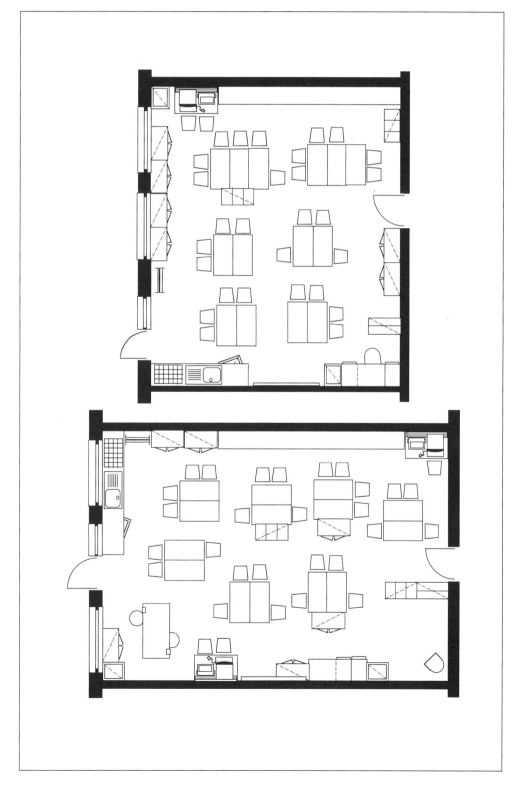

Figure 3.1: basic teaching area in the form of enclosed classrooms of two sizes: at 1.8m² (top) and 2.1m² per pupil (bottom).

resources and facilities. However, if the 'ownership' of the shared area is unclear, each class territory can be perceived as being smaller (even if that is not physically the case). This may result in a greater demand for timetabled supplementary areas such as a studio or a small group room. An example of a pair of semi-open bases is illustrated in figure 3.2.

3.16 The type and character of spaces may vary across the age range. Reception pupils may benefit from a more secure and supportive environment, with space for role play and construction. More independent learning is expected of older pupils, who may use facilities beyond their class base.

Figure 3.2: basic teaching area in the form of semi-open class bases with shared teaching area.

Timetabled Supplementary Areas

3.17 It is often advantageous to accommodate specialised facilities in supplementary areas, such as practical areas, studios

Semi-open Base Example (figure 3.2):
The two class bases illustrated are 35 to 40m², each with space for 30 pupils and their teacher in whole class work. They open onto a shared teaching area which accommodates sinks and other resources including computers. Individual pupils or small groups can carry out activities here which would be inappropriate in their class bases. The two teachers can assist in the supervision of each other's classes and have opportunities for joint working.

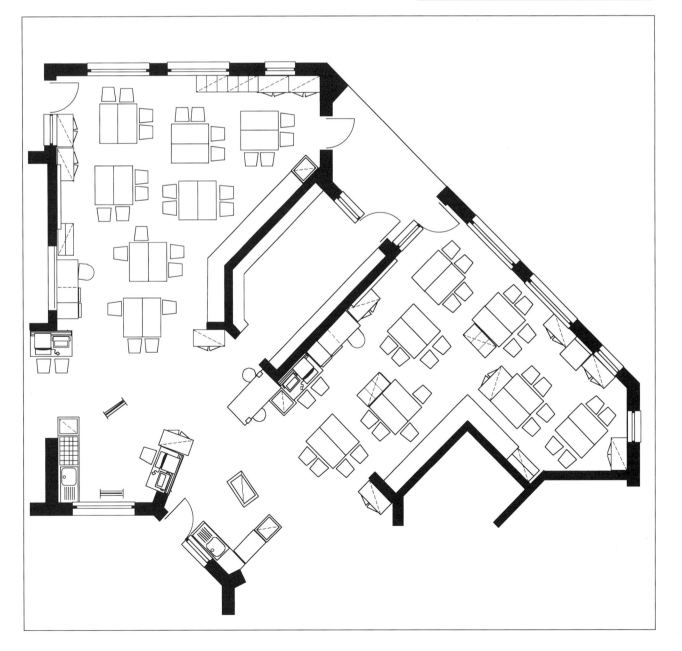

or small group rooms. These are discrete spaces or bays used and supervised independently of the class areas (otherwise they count as basic teaching area). As they will be shared by a number of classes, they will need to be booked or timetabled, like the hall. In larger primary schools and where the number of such timetabled spaces is more critical, the scheduling described in section 2 can ensure maximum efficiency in their provision and use.

Specialist Practical Areas

3.18 Specialist practical areas provide opportunities for sustained and often large-scale work in a safe environment, usually for groups of six to eight pupils. They can be very useful, especially for years 5 and 6, whether in the form of open bays or enclosed rooms.

3.19 Specialist practical areas can be equipped for different activities, from scientific experiment and control technology to cookery and ceramics. They are likely to have a higher level of servicing, special tools and equipment, and more robust furniture and finishes than basic teaching areas. They may also need special storage facilities. Health and safety considerations should be paramount and close supervision is normally required. Figure 3.3 illustrates two types of specialist practical area.

3.20 Figure 3.4 shows a range of areas for practical activities. A specialist practical area can be used for groups of various sizes and occasionally for whole-class activities. The availability of staff or adult helpers is likely to be a major determinant in the provision of specialist practical areas.

3.21 A separate enclosed room has advantages for subjects like food or ceramics. However, these rooms may be under-used and are often too specialised to be used for other activities, so it is useful to determine the likely frequency of use and group size (see appendix 1).

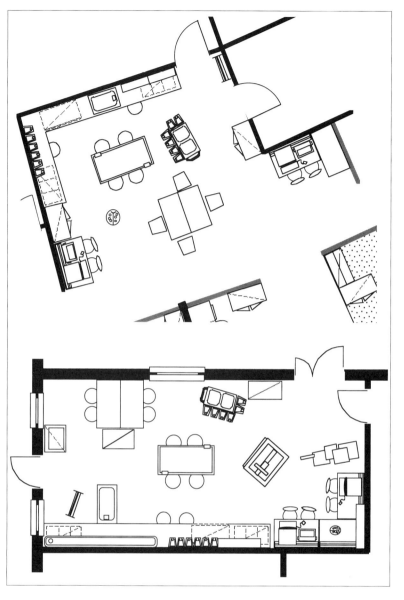

Figure 3.3: two examples of specialist practical areas, as described in paragraph 3.18.

Specialist Practical Area Example (figure 3.3):
Two types of mixed use practical area are illustrated above. The smaller is a 20m² bay, with facilities for up to eight pupils to do a reasonable range of practical activities including:
- wet modelling or painting;
- dry construction on tables, a worktop, a workbench or the floor;
- simple experiments;
- control technology.

Provided that the bay is overlooked by an adjacent class base, it can be supervised by a class teacher who need not stay in this area for the whole session.

The larger is an enclosed room of 36m² with sufficient resources for up to half a class to do an extended range of practical work. In addition to those activities possible in the smaller bay, this room can accommodate:
- water experiments in a sink or trough;
- more ambitious construction;
- 2D and 3D art or graphics.

An adult helper or teacher would need to remain in this room to supervise activities.

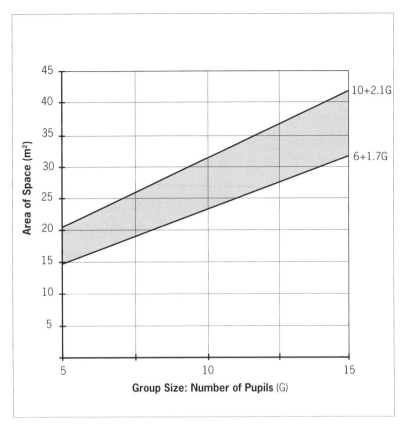

Figure 3.4: graph showing areas for practical areas.

Figure 3.5: graph showing areas for small group rooms and studios.

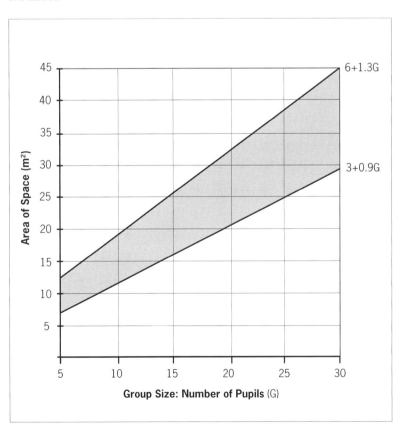

Food

3.22 Ideally, a specialised food preparation and cookery space should be fitted out with kitchen furniture of an appropriate height for the age range. Groups are normally limited to eight pupils for ease of supervision. It is advisable that the food area should be kept separate from other activities, at least in a bay, for hygiene and safety.

Ceramics

3.23 Wherever possible the furniture and equipment in the ceramics area should be of an appropriate size for the age range. Groups usually consist of between six to eight pupils. The ceramics area should be placed in a self-contained bay or separate room to prevent clay, dust and any hazardous substances contaminating other teaching areas.

Small Group Room

3.24 A small group room can be used for withdrawing individual pupils or small groups from a class for remedial or enrichment work. The room can be used for quiet or noisy work, independent study, music composition, drama or for special needs. Figure 3.5 shows the recommended range of areas for different sized groups.

Studio

3.25 Music, drama, movement and dance benefit from a specialised environment with acoustic isolation, 'dim-out' and some stage blocks. Pupils should also have easy access to audio-visual and music resources which can be on trolleys or out on permanent display.

3.26 In smaller primary schools, a hall may meet these conditions as long as there is suitable storage space. In larger schools a studio can be an effective supplement to the hall.

3.27 Figure 3.5 shows the recommended range of areas. Although some activities can be done in smaller groups, a studio may be useful for activities involving the whole class (a group size of 30 in the graph), especially if class bases are unsuitable for music or drama.

Hall

3.28 Most primary schools have a hall for large group activities such as assembly and PE, some drama, music, and often for dining as well. Ideally, the hall should have a piano or keyboard and a full range of PE apparatus, both fixed and loose. There should also be space for loose apparatus to be left out during the day and separate but adjoining stores for loose PE equipment, and for tables and chairs if the hall is used for dining.

3.29 The needs of a class of 30 all doing PE suggests a minimum area of 120m² for infants and 140m² for juniors, excluding storage. If the hall is used for dance and movement without using apparatus, 80m² is acceptable. In some smaller schools these areas may not be available. In such cases, some compromises can be made through careful management of group sizes and activities.

Assembly

3.30 To accommodate assemblies of more than about 400 pupils, the hall has to be larger than the minimum sizes suggested for infants PE. Although the legal requirement for regular collective acts of worship does not require the whole school to

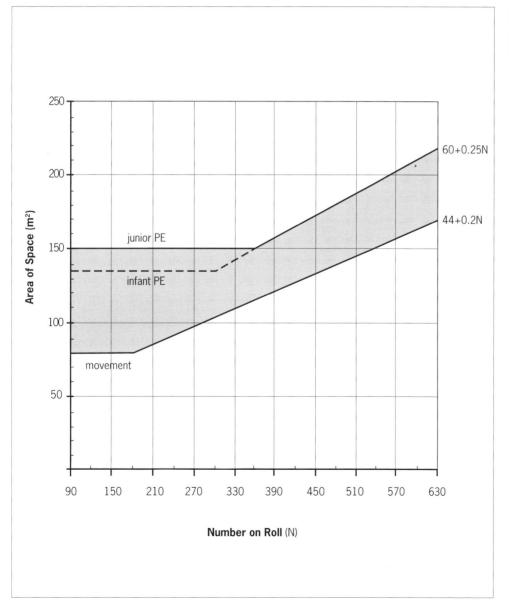

Figure 3.6: graph showing areas for a hall, including PE and assembly.

assemble together, these are occasions that many schools value and this must be evaluated against other priorities. Figure 3.6 illustrates the area for halls which allow the whole school to meet together.

Larger Schools

3.31 In larger schools a single hall may not be sufficient to meet the demands of the curriculum. The exact accommodation requirements depend on factors such as the length of the school day, dining arrangements, timetabling of teaching sessions, and whether music and performance can be accommodated elsewhere. Section 2 and appendix 1 show how to analyse the relevant factors to determine the best distribution of accommodation. A second large space such as a studio may be required and, in exceptional cases, a third space might be needed, large enough for dance and movement (but not for gymnastics with apparatus). The area of a second or third large space must come within the allowance for teaching area and gross area described in section 1.

Other Supplementary Areas

3.32 In most primary schools there is a range of spaces that, by their nature as a resource for the whole school, are not normally timetabled. These include areas for resources and special educational needs, withdrawal spaces and the library. Nevertheless the spaces are counted in the total teaching area for the school and they should be planned for maximum use.

Library

3.33 In some primary schools, library stock is dispersed into book corners in class bases. However, there are many benefits in having a whole school library. It is accessible to all age groups and can contain a much broader spectrum of materials including reference and other books, audio materials, CD-ROM and additional information technology (IT) equipment. The library should also have space for display and tables and chairs for reading, listening and quiet work. Figure 3.7 illustrates the range of areas recommended for different sized schools.

3.34 The location of the library is important. Ideally it should be positioned centrally within the site, easily accessible to the whole school. An attractive and inviting design will stimulate pupils to use it for study and pleasure, and informal use can be encouraged if it is not fully enclosed. Supervision and security are easier to manage if the library is overlooked by teaching or administration spaces.

Resource Areas

3.35 Subjects such as science, technology, art and music require a wide variety of equipment, tools and materials. A designated resource bay or room is often a practical design solution. It is much more than just a storage zone, offering a stimulating display of resources available for pupils' use. Resource areas are best situated adjacent to general teaching or practical areas.

Figure 3.7: graph showing areas for libraries.

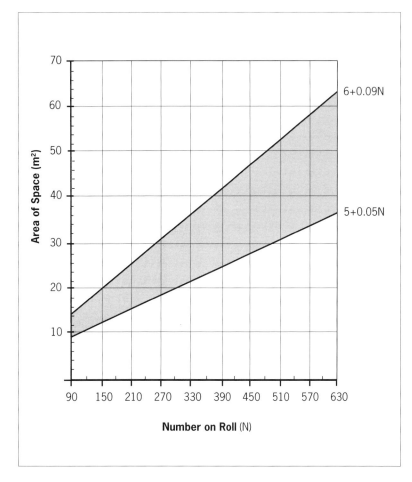

SEN Support Base[1]

3.36 Although pupils with special educational needs (SEN) are usually taught in ordinary classes, with help from special support teachers or assistants, schools providing for a number of pupils with SEN may have a support base for small group or individual teaching. It can include special resources, for use in the base and throughout the school, and space for records and staff preparation. It should be located near other teaching areas, readily accessible to all pupils. It may also function as a small group room.

Non-Teaching Areas

3.37 Provision of the total non-teaching area is determined by the size of school and the number of staff. The paragraphs below describe the range of non-teaching accommodation that may be needed (except where noted otherwise). They are described by function rather than room type as some rooms may be used for several purposes. Different arrangements and combinations are possible.

3.38 Minimum areas are not specified but a typical proportion of non-teaching area that these functions might occupy is listed in figure 3.8. The table relates to an average sized school[2] and the elements of non-teaching area are shown as a percentage of the overall non-teaching area and also as a percentage of the gross area of the school. Additional provision may be needed for pupils with special needs. Clients and designers will be in the best position to assess their own priorities, and to strike a balance within the available gross area.

Staff and Administration Accommodation

3.39 The range of accommodation might be expected to occupy approximately 10% to 12% of the total non-teaching area and includes:

- headteacher's office;
- senior staff office;
- reception;
- school administration office(s);
- staff social space;
- staff work space;
- office resources such as reprographic facilities;
- staff lavatories, showers and changing facilities.

3.40 In a school with more than 120 pupils, the *headteacher* must have an office, for meetings and interviews which may be confidential. Only larger schools justify a deputy head's or senior staff room and then this room might double up for other uses such as medical inspection. Sometimes these offices are used for the storage of records, valuables and cash in which case some form of additional security may be necessary.

3.41 School administration areas are often linked with the reception. If this is the case, the space should be accessible to staff, pupils and visitors whilst at the same time providing a secure area for the school records.

3.42 Staff social spaces should ideally provide an area for refreshment and relaxation and also some group work. There should also be an area within the room or elsewhere in the school that is designated a *staff work space*. This is needed for preparation, assessment and recording work. Except for larger schools, these functions will normally be combined in one space.

3.43 The *reprographic facilities* may be placed in the school administration area. However, this can be noisy and disruptive and it may be preferable to use a separate small room or store area for reprographics. In larger schools it may be necessary to have separate office resource areas for school administration and teaching preparation.

3.44 Staff lavatories and washbasins should be separate from those provided for the pupils[3]. A lavatory for the disabled can be unisex and may be used by adults and pupils (if it is in a self-contained room off a circulation space).

Notes
1: See building bulletin 61: Designing for children with special educational needs: ordinary schools, HMSO 1984, now out of print but available from HMSO Books (Photocopies Section).
2: An average sized primary school of seven classes and about 210 pupils.
3: Workplace Health, Safety and Welfare Regulations (1992) Approved Code of Practice gives guidance on the minimum provision of sanitary conveniences for employees.

type of space	proportion of total non-teaching area	proportion of gross area
staff and admin. accommodation	10% - 12%	4.0% - 5.2%
pupils' storage/washrooms	17% - 19%	6.8% - 8.2%
teaching storage	12% - 13%	4.8% - 5.6%
catering facilities	7% - 10%	2.8% - 4.3%
ancillary spaces	9% - 10%	3.6% - 4.3%
circulation and partitions	39% - 42%	15.6% - 18.1%
total non-teaching area		**40% - 43%**

Figure 3.8: proportional breakdown of non-teaching areas.

Pupils' Storage/Washrooms

3.45 These could take up approximately 17% to 19% of the non-teaching area and include:

- pupils' lavatories;
- pupils' coat store(s);
- lunch box storage;
- pupils' changing room(s);

3.46 The School Premises Regulations contain specific requirements for pupils' sanitary facilities. For all except nursery pupils, one sanitary fitting must be provided for every twenty pupils. Unisex lavatories are generally regarded as acceptable for infant or nursery pupils only. Pupils aged eight and older must have access to separate male and female washrooms. The number of wash-hand basins should equal the number of sanitary fittings in each washroom.

3.47 Pupils' coat store(s) need to be provided for the storage and drying of outdoor clothing and bags[1]. Their position, heating and ventilation need careful integration within the school environment. If the coat stores are dispersed to class bases the space required should not be taken out of the teaching area. If pupils are bringing lunch boxes to school they may need to be stored in a controlled area (at the correct temperature and away from potential contamination).

3.48 Pupils' changing rooms are not a statutory requirement but it may be considered useful to have at least one which is large enough for half a class.

Teaching Storage

3.49 The total teaching storage might be expected to occupy approximately 12% to 13% of the non-teaching area. It includes:

- class storage;
- secure stores;
- PE storage.

3.50 The *class storage* is for teaching materials and equipment, and for items of stock that need to be stored in a class base or shared area for ease of access by pupils (see paragraph 3.2). The *secure stores* might be used for stock, bulk or valuable items which are only needed occasionally. The *PE storage* is necessary for equipment and apparatus for indoor and outdoor use.

Catering Facilities

3.51 The scale of catering operations, the number of staff employed and the area required can vary enormously. The percentages in figure 3.8 do not allow for a full production kitchen. If the school is providing cooked meals, heated and served on the premises, the catering facilities may occupy approximately 9% of non-teaching area and might include:

- a servery or scullery for serving meals, whether delivered or cooked on the premises, and washing-up afterwards;
- a chair and table store, possibly combined with the servery;
- catering staff facilities, such as a lavatory, office and rest room.

3.52 The majority of primary schools no longer operate full size production kitchens and the gross area guidelines assume an area of about 35m[2] for a sophisticated servery only.

Ancillary spaces

3.53 The number and range of ancillary spaces can vary according to how the school is organised and managed. These spaces can occupy between 9% and 10% of non-teaching area and include:

- a medical room;
- a caretaker's office, for administration and maintenance (if required);
- maintenance storage, cleaner's and caretaker's stores (depending on arrangement);
- plant and fuel store (if required).

3.54 All schools must have a *medical room*, used for medical and dental inspection and for pupils and staff who feel unwell. It must have a wash-hand basin and be located close to a suitable lavatory and should be easily supervised. It need not be used solely as a medical room, but should be readily available for use as such (as in paragraph 3.40).

Circulation and Partitions

3.55 The *circulation* routes need to be wide enough for fire safety and for pupils to move easily throughout the school between sessions. The area designated for circulation cannot normally be used for educational activities, although teaching space may open directly onto it (subject to fire and safety precautions). But it does add to capital and recurrent costs, so wherever possible designers should try to minimise it.

3.56 Space for *partitions* comprises the floor area occupied by partition walls but excludes the area occupied by external walls. The proportion of non-teaching area taken up by circulation routes and partitions can be between 39% and 42%.

Nursery Provision

3.57 Nursery provision in schools takes a variety of forms. It can be in the form of:

- self-contained schools with one or more classes;
- a nursery class or classes within a primary school;
- a unit of one or more classes on a primary school site but in physically separate buildings.

Maintained nurseries are also sometimes combined with day care and other facilities in integrated nurseries or early years centres.

3.58 The main planning ingredients are described below and illustrated by the plan in figure 3.9. At the time of writing, more detailed guidance is being prepared, but reference may be made to existing DfEE publications[2].

Teaching Area

3.59 The bulk of the teaching or play area for nursery pupils is usually the playroom, including space for wet and messy activities. The teaching area may also include a small group room.

Teaching/Playroom

3.60 Common to all forms of provision is a teaching/playroom. A well designed one will include informally demarcated zones for the following activities:

- table work by at least a third of the children at any one time;
- role play;
- musical activities;
- wet and messy work;
- quiet activity;
- large-scale construction;
- use of IT.

3.61 It is good practice to provide sufficient unobstructed space for many activities to take place on the floor. Mobile furniture and partitions, carpets and

Notes
1: See A&B Paper number 15, Lockers and Secure Storage, DES 1990 and Building Bulletin 58: Storage of Pupils' Personal Belongings, HMSO 1980.
2: Building Bulletin 56: Nursery Education in converted space, HMSO 1976, now out of print but available from HMSO Books (Photocopies Section). Design Note 1: Building for Nursery Education, DfEE 1968. Broadsheet 1: Nursery education: low cost adaptation of spare space in primary schools, DES 1980.

waterproof floor finishes help to define these zones. Within the space a range of books, games, toys, equipment, instruments, materials and other resources would be on display and accessible to the children. If provided, mid-day meals will usually be eaten here too. The activities described can be accommodated within an area of 2.3m² per pupil. A higher area per pupil may be preferable in some circumstances, but this is subject to ongoing research.

Wet and Messy Area

3.62 An important part of teaching/play activity involves water, sand and other messy work, preferably on a waterproof floor finish. Ideally, hot and cold water should be to hand. The most versatile water source is a deep sink with a drainer, projecting to allow children to gather around. It can be useful for this space to link directly to the outdoor play area.

Small Group Room

3.63 A separate, smaller room is usually an asset. This can be used for a number of the activities already described, including quiet activities such as story-telling or withdrawal of individual children, where some acoustic or visual separation is an advantage. Characteristics of the space include soft finishes, bean-bags and comfortable furniture. To gain the best value from a separate room, it should be used as much as possible, which requires either visual supervision from the main space or the availability of additional adults.

3.64 A room of this kind counts as part of the total teaching area. If community or other uses are expected, other rooms may be needed.

Outdoor Area

3.65 Access to adequate outdoor teaching and play areas is important, as many activities go on in the fresh air for a large part of the year. A covered outdoor area opening off the teaching/playroom allows outdoor activities to continue even in inclement weather. Such an outdoor space should be secure and within the area guidance given in paragraph 5.33.

Non-Teaching Areas

3.66 A successful teaching/playroom will depend on the careful planning of some essential support areas, including:

- an entrance/coats area;
- pupils' lavatories;
- wash-down facilities;
- storage.

3.67 Certain non-teaching accommodation is required by the School Premises Regulations, either in the nursery itself or in the school to which it is attached. This includes:

- the headteacher's room (in schools with more than 120 pupils);
- staff accommodation;
- staff washroom(s);
- a food preparation area;
- a medical room;

Other ancillary spaces may also be required. An integrated nursery or early years centre may also require accommodation for younger children, day care and other facilities.

Staff and Administration Accommodation

3.68 All nurseries should have a base with space for administration, record-keeping and interviews. This will often be the headteacher's room, but may be combined with a staff or other room closer to the nursery. If so, it is useful if the room overlooks the main entrance to the nursery.

3.69 A staff room and lavatories, for visitors as well as staff, are required. They may be shared with the main school.

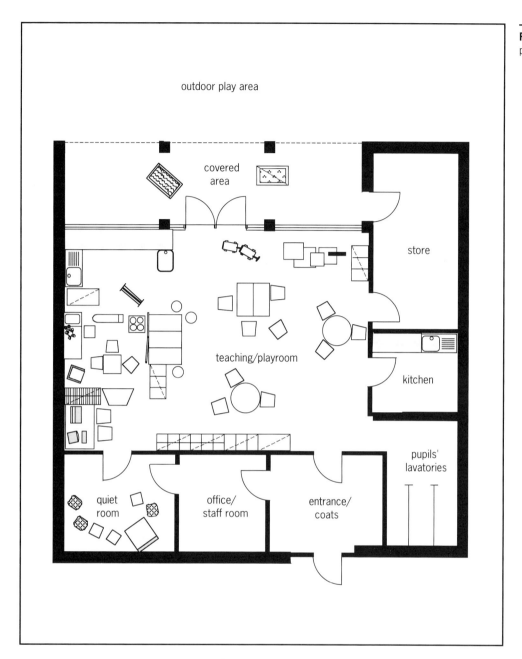

Figure 3.9: example of a 26 place nursery unit.

Nursery Example (figure 3.9):

The example above is a 26 place nursery attached to a primary school. The headteacher's room, adult lavatories, kitchen and medical room are in the main building near by. Approximately 60% of the gross area of the unit is teaching area in this example.

The entrance to the nursery leads into a lobby with room for coats and buggies, overlooked by a small staff room which is also used as an office. The pupils' lavatories are close by. The playroom is 51m² and an adjacent quiet room is 9m², amounting to a total teaching area of 2.3m² per pupil. Furniture in the playroom is

informally laid out to allow table work, role play, music, wet and messy work, dry construction and work with computers, with clear areas for activities on the floor. The wet area has a worktop and sink at child height and a projecting sink for experiments. Both wet and dry practical areas open directly into a large covered area, beyond which is the outdoor play area.

Most of the furniture is mobile and a range of books, toys and other resources is displayed for easy access by pupils. Items which need to be stored securely can be locked in the store, accessible from both outside and inside. A small kitchen provides facilities for snacks and a washing machine.

Pupils' Storage/Washrooms

3.70 One sanitary fitting must be provided for every ten nursery pupils. Lavatories are usually grouped within easy reach of the teaching/play space and can be unisex. The design of the partitions should allow some privacy for children and space for adults to give assistance, yet permit adequate supervision.

Wash-down Facilities

3.71 A deep sink, bath or shower tray is required to clean accidentally soiled pupils or clothes (generally one for every 40 nursery pupils). A laundry or utility room with a washing machine may also be useful for this and other purposes. A changing table or height-adjustable trolley should be considered for children with special needs.

Entrance/Coats Area

3.72 The procedure of taking off and storing outdoor garments, lunch boxes and personal belongings is a useful discipline to learn. Space may be needed for parents to assist, which often includes the need to fold, unfold or store prams or buggies without obstructing the fire escape route. A gate between the entrance door and the street will increase safety.

Teaching Storage

3.73 Storage for toys, materials and equipment is needed. A store for apparatus and wheeled items directly accessible to the outside play space is also desirable.

Food Preparation

3.74 All nurseries need provision for the preparation of snacks, and some will offer mid-day meals. Facilities may range from a worktop with ring or power-point to a fully equipped kitchen, which may be elsewhere in the main school. Safety for children is essential, and physical separation, whether in another room or by means of low barriers, is always advisable.

Ancillary Spaces

3.75 These may include spaces for a cleaner or caretaker, plant, circulation and other space as appropriate (as described in paragraphs 3.53 to 3.56).

3.76 Every nursery should have access to a medical room (see paragraph 3.54). It may be shared with the main school and need not be close to the nursery, provided it can be adequately supervised. However, if it is situated close to the nursery, it may have additional value as a base for visiting specialists in the assessment of special needs.

Section 4: Secondary Accommodation Guidelines

This section provides guidance on teaching and non-teaching spaces in secondary schools. Teaching accommodation comprises three types of **timetabled teaching area**: general teaching, practical areas and physical education (PE) spaces, plus **non-timetabled teaching areas**, including supplementary spaces. **Non-teaching areas** are covered at the end of this section.

Figure 4.1: illustration of possible demand for rooms according to activity in secondary schools.
* includes space requirements for both vocational and core skills.

Timetabled Teaching Areas

4.1 The secondary school curriculum is normally taught in distinct subjects, using a variety of timetabled rooms. These rooms or spaces tend to be used primarily for one subject. Hence the secondary area guidelines are described under subject headings in the following text. However, it is important to note that some rooms may be used by more than one subject, for example an English room may be used for teaching history or a specialist space may be used for music, drama and modern foreign languages. Figure 4.1 illustrates the types of spaces commonly used for different subjects. Almost half the subjects taught in secondary schools are 'general teaching' requiring standard classrooms. Other subjects need specialist practical areas or sports facilities.

4.2 The following considerations are appropriate to all secondary teaching spaces.

Room Shape

4.3 A space which is too narrow may restrict the range of activities and the possible furniture layouts. This is particularly true for practical spaces where there may be large items of equipment. Most subjects at secondary level include some formal teaching. The proportions of a space should give good sight lines to the whiteboard. The most useful shape of space is normally a square or a rectangle of a proportion of around 4:5.

Suites and Links

4.4 Most secondary schools are organised into departments or faculties. Arranging rooms into departmental suites enables resources, equipment, store rooms and supplementary teaching spaces to be

timetabled spaces / subjects	standard classroom	large classroom	GNVQ/ business room	IT room	science laboratory	multi-materials workshop	pneumatics, electronics and control technology	food room	textiles or graphics space	2D and 3D art spaces	music room	music recital room or drama/media studio	PE spaces or main hall
English	○	○		○								○	
mathematics	○	○		○									
humanities	○	○		○	○								
modern foreign languages	○	○		○								○	
RE/PSE	○	○											
IT courses			○	○									
business studies		○	○										
science	○			○	○	○	○						
design and technology				○		○	○	○	○				
art									○	○			
music		○									○	○	○
drama		○										○	○
PE/ sports studies	○			○								○	○
GNVQ business*		○	○	○									
GNVQ art and design*	○		○	○		○			○	○			
GNVQ leisure and tourism*	○		○	○									○

shared. Departments which may be linked, such as design and technology and art, should also be positioned close to each other. Whole school resource areas like the library resource centre should be easily accessible from all departments.

Servicing

4.5 All teaching spaces should ideally be serviced for the use of computers and audio-visual teaching aids. Computer networks may be provided as local area networks or whole school networks. Some practical spaces, such as science laboratories and multi-materials workshops, have greater servicing requirements. The method of distribution of all services should allow the appropriate range of activities to take place conveniently and safely.

Room Sizes

4.6 Under each of the following subject headings, the likely required size of each room type is represented by a zone (A -

H) on a graph. These zones illustrate a range of areas for different group sizes[1]. The range reflects the variety of situations that can be found in schools.

4.7 The notes for each type of space describe the effect that different activities may have on area. General factors which may affect the size of space include:

- the type of courses offered;
- the extent of supplementary areas (a space may be smaller if common activities take place in a shared resource area);
- any storage in the room.

4.8 Some local storage is allowed for within the spaces (for example below a perimeter worktop), but a nearby store room may also be required. Additional storage areas are given under each subject, along with any resource or supplementary areas that may be needed.

4.9 Areas are given on the graphs for 11-16 groups (G) and post-16 groups (g). The area for the 16-18 groups are a proportion of the 11-16 area (g = G ÷ 1.5) to give a broad estimate only. Ideally all spaces should be big enough for the largest group size likely to use the space, to allow greater flexibility in allocating groups to rooms.

General Teaching Areas

4.10 General teaching subjects include:

- English;
- mathematics;
- modern foreign languages;
- humanities (history and geography);
- religious education;
- personal and social education (PSE);
- general studies.

Such subjects usually require ordinary classrooms, although some may occasionally need practical spaces. Conversely, some subjects which are usually taught in practical or PE spaces can also be timetabled for certain lessons in general teaching rooms: for example, theory lessons in PE, science or elements of vocational courses. General teaching spaces are often used for class registration or social activities.

Figure 4.2: graph showing areas for standard and large general teaching classrooms, zones A and B. In the formulae, G is the KS3 or KS4 group size. The sixth form group size (g) is shown on the lower line of the horizontal axis (g = G ÷ 1.5).

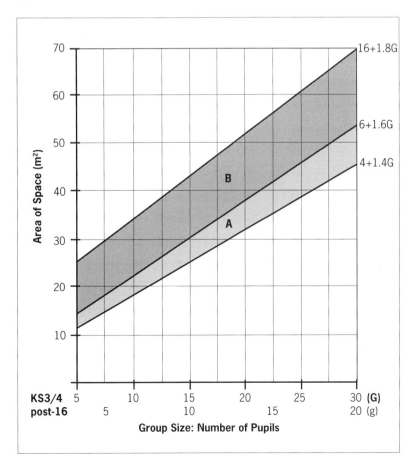

Activities

4.11 The range of activities in general teaching rooms includes: whole class teaching and small group discussion, reading, writing, role playing, and the use of computers and audio-visual equipment. There may also be a limited amount of other activities such as model making, for example in mathematics and geography.

4.12 Although subjects such as geography and languages may require a degree of specialisation, general teaching rooms tend to be used for more than one subject. For example, an English room might be used to teach mathematics for two periods per week. This means that general teaching rooms can have an average frequency of use as high as 90%.

Types of Spaces

4.13 Most of the activities that take place within general teaching can be catered for by two sizes of space: *standard* and *large*.

Standard Classroom (Zone A)

4.14 **Zone A** in figure 4.2 shows a range of sizes of rooms for standard general teaching spaces. The areas represented are suitable for all general teaching subjects. The range depends on the level of furniture and equipment in the room. For example, if computer workstations are required, the higher end of the range may be more appropriate. The lower group size figures from this graph can also be used for calculating areas of small rooms used by small groups.

Large Classroom (Zone B)

4.15 On occasions it may be necessary to have a larger space, for example where mathematics requires more computers, or geography needs space for map work and model making. Activities such as role play in modern foreign languages sometimes warrant a larger space. One such space for every four to five standard spaces may be a good ratio. **Zone B** in figure 4.2 shows the area that might be appropriate.

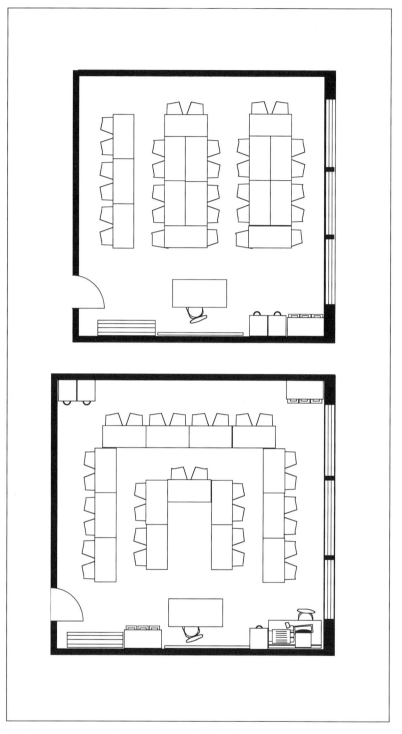

General Teaching Examples (figure 4.3):
Two standard general teaching spaces for 30 KS3 or 4 pupils are illustrated above, from the upper and lower ends of zone A. They both have space for a coat and bag rack beside the door and some teaching storage, but the larger classroom of 54m² has space for a computer workstation and a more flexible furniture layout. The smaller classroom, at 48m², has less storage and equipment. It may support some activities less easily and limit possible furniture arrangements.

Figure 4.3: examples of two sizes of standard general teaching classroom.

Note
1: The likely group size should be determined from a curriculum analysis or discussion with the school.

4.16 Untimetabled supplementary teaching areas may be required in suites of general teaching accommodation, for instance a room for use by a foreign language assistant to work with small groups or for careers advice. Small clusters of computers may be in a shared area rather than in the class (paragraph 4.74).

4.17 Some storage in furniture within the room is allowed for in zones A and B. Additional storage space may be necessary either within the classroom or in a separate walk-in store. A floor area of approximately $0.08m^2$ to $0.09m^2$ per workplace usually provides enough full height storage for general teaching.

Practical Areas

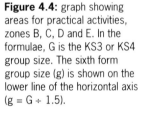

Figure 4.4: graph showing areas for practical activities, zones B, C, D and E. In the formulae, G is the KS3 or KS4 group size. The sixth form group size (g) is shown on the lower line of the horizontal axis (g = G ÷ 1.5).

4.18 Practical spaces house specialist activities and equipment and tend to be specific to a number of practically-based subjects, including science, design and technology, art, music and drama. Vocational courses such as GNVQs may involve units that are taught in a number of different specialist areas, which may then be shared with other subjects and age groups. This can be arranged within the timetable (as in appendix 3). Similarly, some National Curriculum subjects such as design and technology will require a number of types of specialist space.

4.19 The specialist nature of practical spaces means that they are less interchangeable than general teaching spaces. For example, the practical elements of food technology cannot be taught in a music room. However, a frequency of use of between 75% and 85% of the timetabled week can normally be achieved. To ensure this efficiency of use and to offer a wide range of spaces, two subjects or activities may well be carried out at different times in one practical space, for example textiles and graphics. This has to be planned in advance so that the room can be properly equipped. Section 2 and appendix 2 give examples of how this can be carried out.

Activities

4.20 In addition to the particular specialisms, all practical spaces need to cater for activities such as whole class briefing and evaluation, small group discussion and using computers.

Types of Spaces

4.21 As practical spaces are often predominantly used by one subject, the types listed below have been grouped under their likely subjects. The area needed for particular group sizes varies according to the scale of specialist activity and size of the furniture and equipment. The zones (C - H) in figures 4.4, 4.7 and 4.8 reflect these differences. The most typical area range for each type of space is given in brackets after its sub-heading. The zone above and below that 'standard' will cover a broader range of areas which might be applicable in some situations. For example, a room in the zone below may be appropriate if it is one space amongst a group of larger spaces. The zone above

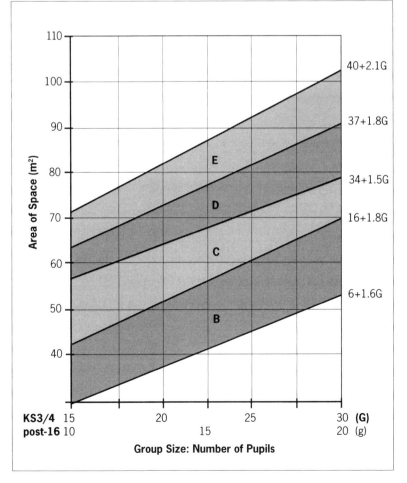

may be referred to where a number of different activities take place in the same area. A space used by sixth form pupils alongside younger pupils will tend to come from the upper part of the ranges or the zone above.

4.22 There are additional untimetabled teaching spaces and teaching storage areas associated with these practical subjects. These are mentioned under each subject heading and are summarised at the end of this section. Area ranges for storage requirements are given for most subjects. As with general teaching, these figures are in addition to any local storage included in the teaching spaces. Since a basic level of storage is always required, the upper end of the range may be more suitable in a small department. The lower figure may be appropriate where storage and/or preparation is concentrated into one space.

Business Studies and Vocational Courses

4.23 Business courses involve a variety of activities including group discussions, presentations and using IT. Some of these activities can be carried out in a large general teaching room (**zone B**), but an IT room or business studies room with specialist facilities may be more suitable for the majority of the time.

4.24 Pupils studying vocational subjects need frequent access to computers, both in timetabled work and in their own study time. These may be available in easily accessed resource areas, or in classrooms used for 'core skills' such as numeracy, but will often be provided in a 'base room', similar to or the same as that used by business studies. GNVQ courses such as health and social care or business will require this type of room for most of the curriculum. Courses such as manufacturing will need it for less time, as most teaching will take place in more specialist practical spaces such as a workshop. An example of an 11-18 school teaching a number of vocational courses is shown in appendix 3.

GNVQ/Business Room   **(zone C)**

4.25 A practical space with an area in **zone C** allows for a range of activities, including using computers, related to GNVQ and business courses. Certain activities such as simulating working in an office may require an area in **zone D-E**.

4.26 GNVQ and business courses may use untimetabled spaces such as IT resource areas and seminar rooms. The additional teaching storage requirements may vary between $0.3m^2$ and $0.4m^2$ per workplace.

Information Technology[1]

4.27 Whilst information technology (IT) equipment is used across the curriculum, it may be taught as a course to allow pupils in lower years to learn basic skills. It may also be taught as an examination subject, although the demand for this is decreasing. IT facilities may also be placed in small untimetabled clusters around the school (paragraphs 4.68 and 4.74).

IT Room   **(zone C - D)**

4.28 IT rooms are usually available to be booked on an ad-hoc basis for a variety of subjects, such as English or humanities, as well as being partially timetabled for IT courses. The size of an IT room depends on the number of machines and supporting hardware, such as printers, and the area allowed for pupils to carry out associated graphic and written work. It is likely to be in **zone C-D**. If the room is shared with business studies or is used for desk top publishing, it may be more appropriate to refer to **zone E**.

4.29 Additional teaching storage requirements may range from $0.3m^2$ to $0.4m^2$ per workplace.

Science[2]

4.30 Science involves a high percentage of experimental work, particularly at KS3 and KS4, requiring regular access to a range of services. Some science lessons can take place in general teaching rooms, but most will require a laboratory.

Notes
1: The video and leaflet 'Making IT Fit' (DfEE 1995) provide guidance on accommodating computers in schools.
2: Building Bulletin 80: Science Accommodation in Secondary Schools, HMSO 1995, covers detailed design.

Science Laboratory (zone D)

4.31 Activities that need to be accommodated in science include small groups of pupils carrying out experiments and teacher demonstration. The size of any laboratory will depend on the furniture system to be used and the group size. An area from the middle of **zone D** allows for a typical range of activities. Laboratories in the lower area of **zone C** might have a limit on the range of activities and the choice of furniture they can accommodate. This size would not be recommended for all the laboratories in a suite. A laboratory in the upper part of **zone D** could accommodate specialist sixth form equipment and long-term experiments alongside general science provision.

4.32 Untimetabled rooms which supplement the main teaching areas may include:

- an animal room;
- a microbiology room;
- a greenhouse;
- an outdoor project area;
- a darkroom;
- an IT resource/display area.

4.33 Preparation areas are needed in science for cleaning and sorting laboratory equipment as well as storing materials. A floor area of between 0.4m² and 0.5m² per workplace should provide sufficient overall storage and preparation area. The lower figure will be appropriate when one

Science Examples (figure 4.5):
The examples below illustrate how the area ranges in zones C and D can apply to a science laboratory. Both spaces provide for 30 pupils to take part in practical and non-practical activities. The larger space is from the upper part of zone D. Additional benching at the back of the room allows sixth formers to set up and monitor long-term experiments. There are two computer workstations, a position for a mobile fume cupboard and a rack for pupils coats and bags. The smaller space is at the lower end of zone C. There is no fume cupboard and no additional table for a computer. Coats and bags are stored outside the room.

Figure 4.5: two sizes of science laboratory at opposite ends of zone D.

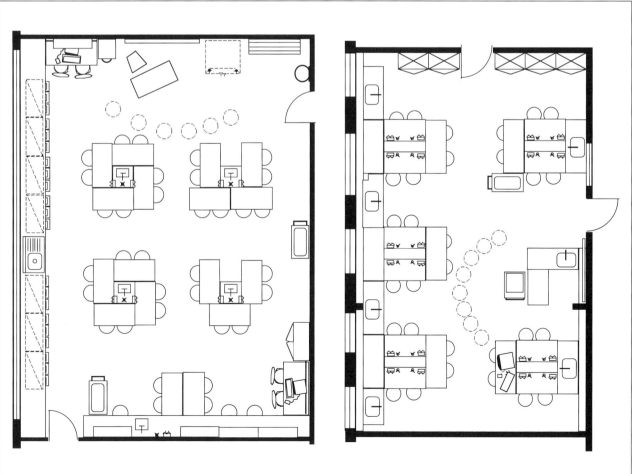

central preparation room is used. This is economical and more convenient for technicians than several smaller spaces.

Design and Technology[1]

4.34 Design and technology encompasses a range of designing and making activities. These can be divided into the following specialist categories:

- food;

- multi-material construction, including working with wood, metal and plastics;

- control technology including electronics and pneumatics;

- textiles;

- graphics.

Teaching spaces may accommodate one or more of these specialisms. Combinations of these activities are more likely to occur in small schools[2].

4.35 As well as the untimetabled supplementary teaching areas described later under multi-materials and food, a design and technology suite may have a shared IT or design/resource area (see paragraph 4.76), a display area or a sixth form project space within it.

4.36 The number of the different types of timetabled rooms depends very much on the proportion of specialisms done in design and technology in each school. For example, a school may choose not to offer food as a component of design and technology. The following text describes a typical range of specialist spaces. Notes are also included on the most likely combinations of specialist activities.

Food Room (zones G - H)

4.37 Specialist activities in a food room include food preparation, cooking and analysis. The size of space will depend on the number and variety of workstations. An area from the middle of **zone G-H** accommodates a broad range of practical and non-practical facilities. A space from **zone F** may be suitable where two or more food rooms share certain common facilities (see paragraph 4.38). A space

from the upper part of the range in **zone H** can allow a very high ratio of cookers to pupils alongside less practical activities.

4.38 A supplementary teaching areas next to a food room may be used for experiments in food technology or food tasting. Some facilities, such as laundry equipment, could be in an adjacent room, allowing the main room to be smaller. Additional teaching storage for food technology might amount to between $0.5m^2$ and $0.6m^2$ per workplace.

Multi-materials Workshop (zones G - H)

4.39 Multi-materials activities involve the use of floor standing and bench mounted machines as well as hand tools at work benches. IT may be used for both design and manufacture, with computerised machinery sitting alongside manual machinery. The machine provision affects the size of space. Where a single room is used for multi-materials, the upper part of **zone H** may be required. A pair of work-shops, both with an area from the middle of **zone G-H**, can accommodate a broad range of facilities including some graphics. An area in **zone F** may be suitable if, for example, drawing facilities are available in an adjacent area or if there are other larger multi-materials rooms nearby (see figure 4.6). An area above **zone H** may be appropriate where:

- facilities are provided for pneumatics, electronics and control technology (PECT) alongside an adequate range of multi-materials facilities;

- post-16 pupils are working alongside KS3 or KS4 pupils on industrial projects, for instance on vocational courses.

4.40 Supplementary teaching areas adjacent to a multi-materials workshop might include an external project area and a heat treatment bay, located within a defined area near or within the workshop for ease of supervision. An area of between $14m^2$ and $20m^2$ will accommodate a reasonable level of heat treatment equipment. A preparation area of about $30m^2$ to $35m^2$ will accommodate preparation machines and associated materials used by

Notes
1 : Building Bulletin 81: Design and Technology Accommodation in Secondary Schools, HMSO 1996, covers detailed design.
2 : Health and safety issues must be considered when combining activities.

Multi-materials Examples (figure 4.6):
The examples below show a pair of multi-material spaces and a single workshop for up to 20 pupils: the pair of spaces are taken from zone F and the single space from zone H. All three spaces are multi-material but the two smaller workshops act as a pair, each having pedestal machines with a different material bias. The single workshop is assumed to be the only workshop, possibly in a small school. The full range of machines is therefore concentrated into one area and the space is larger than a typical space from zone G-H. Both examples provide a basic level of bench and pedestal machinery suitable for use with metal, wood and plastic, with provision for plastics equipment either on a trolley or bench. There are vices for up to 20 pupils. Multi-benches allow provision for hand tools and CNC machines enable computerised design as well as making. The larger single space has tables for designing and finishing activities, the smaller pair of workshops do not. The smaller spaces might be appropriate where drawing and designing take place in an adjacent resource area, possibly shared with other design and technology subjects. Alternatively, multi-bench tops can be covered and used as tables. In both drawings a heat bay is assumed beyond a low wall indicated in white, similarly a preparation area is assumed through double doors on the opposite wall.

Figure 4.6: examples of two sizes of multi-materials workshops; two at 85m² and one at 109m².

the technician. This area could be combined with the additional storage requirements of multi-materials which range between 0.6m² and 0.8m² per pupil. The preparation area should be located near the multi-materials workshops.

Pneumatics/Electronics and Control Technology (PECT) Room   (zone F)

4.41 Typical activities in PECT include designing, making and experimenting using mechanical, electrical, pneumatic and electronic components and systems. IT will be used both for design and manufacture, perhaps using CAD-CAM machinery[1]. The size of space will be determined partly by the range of activities and the provision of computers and machinery for design and manufacture. An area from **zone F** allows for a range of activities including small scale construction involving hand tools. If these facilities are not required a space from **zone E** could be used.

Textiles Space   (zone E - F)

4.42 Textiles activities in design and technology usually include a range of activities such as sewing, weaving, knitting and materials analysis. The size of the space depends on the extent of the activities and equipment. An area from the upper part of **zone D** may accommodate a limited range of activities but an area from **zone E-F** gives greater flexibility. If the school offers a vocational course in art and design or if textiles activities such as screen printing take place in the same space, an area from the upper part of **zone F** may be suitable (see paragraph 4.46). The additional teaching storage requirement for textiles falls between 0.4m² and 0.5m² per workplace.

Graphics Space   (zone E)

4.43 A specialist graphics room may also be used for design and technology particularly in large departments, although graphics often takes place in the other rooms already listed. An area from **zone E** allows for a broad range of activities including modelling and CAD[1].

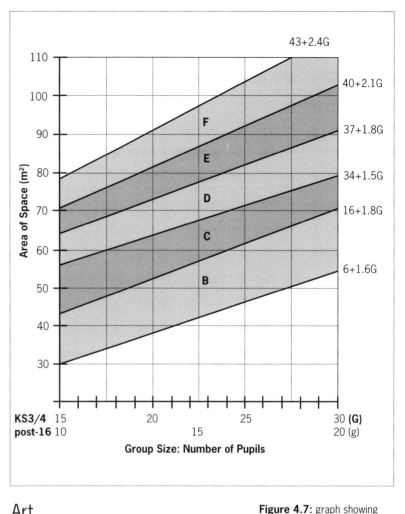

Art

4.44 Art involves work in two and three dimensions in a variety of media including paint, ceramics and textiles.

General Art Room   (zone D - E)

4.45 An area around the middle of **zone D-E** accommodates a broad range of mainly two dimensional art activities, such as drawing, painting and graphics. An area from the lower part of **zone D** may be suitable but it is likely to limit the facilities that can be provided.

3D Art/Textiles Studio   (zone F)

4.46 A space which is equipped for activities such as screen printing onto textiles or three dimensional construction work using plaster or clay may require an area from **zone F**. If facilities for ceramic work are included in a room with other three dimensional art, an area from the middle of **zone F-G** may be appropriate.

Figure 4.7: graph showing area zones B to F. In the formulae, G is the KS3 or KS4 group size. The sixth form group size (g) is shown on the lower line of the horizontal axis (g = G ÷ 1.5).

Note
1: CAD-CAM machinery uses computer aided design (CAD) to control a computer aided machine such as a lathe.

4.47 Supplementary teaching and re-source areas used in art might include:

- an IT resource/display area;

- a darkroom;

- a kiln room;

- a sixth form project room;

- an outdoor project area.

4.48 The additional teaching storage requirements for art fall in the range 0.4m² to 0.5m² per workplace.

Music

4.49 Activities in music include listening, composing and performing. There may also be extended curricular activities ranging from individual instrumental lessons to orchestral concerts. Listening activities usually involve a whole class and may take place in a standard classroom if the acoustic conditions are suitable. For most activities in music a specialist space is preferred, supported by group rooms, allowing pupils to disperse in small groups for some composing activities.

Music Rooms   (zone B or D)

4.50 Composing and performing involve the use of instruments and often take place in small groups. An area from the upper half of **zone B** can accommodate a range of practical activities.

4.51 An area in **zone D** enables extended curricular activities such as recitals or choir rehearsals to take place in addition to the activities already described. This may be useful as one of a group of music spaces. Music may also be performed to an audience in a larger space (see paragraph 4.55).

4.52 The supplementary teaching areas associated with music include group rooms, for instrumental teaching and composition in various small group sizes, and a recording control room. Additional teaching storage requirements for music amount to between 0.3m² and 0.4m² per workplace.

Drama

4.53 Drama activities may be carried out in a large classroom or main hall as part of English, but if drama is taught as an examination subject, a suitably equipped specialist studio is preferable.

Drama/Media Studio   (zone D - E)

4.54 An area in **zone D-E** is suitable for class work in drama with no other audience. A similar space in could be used as a studio for activities in media studies, such as making videos, as audiences need not be accommodated.

Performance Space (zone F and above)

4.55 An area in **zone F** allows for performances to small audiences. Such a space may also be suitable for musical performances or dance activities and may need to have a high ceiling to hang stage lighting and to enhance the acoustics of the room.

4.56 Music, drama and extra-curricular activities such as school plays will require a space to perform in front of a large seated audience. The area needed for will be above zone F, so schools may use:

- a specialist theatre space, normally timetabled for drama;

- the main assembly hall;

- a gymnasium or sports hall, if suitable flooring is provided and the acoustics are acceptable;

- occasionally an outdoor amphitheatre, in fine weather.

4.57 The areas for assembly described in paragraph 4.65 will be needed for a space where a large audience is to be accommodated. An area for the stage and wings will need to be added, similar to that of a drama studio. If the stage area is acoustically discrete, it may be used as a separate teaching space. If the space is also used for PE activities, the areas described in paragraph 4.59 and figure 4.9 can be used to estimate the size of hall required.

4.58 Spaces used for performances may have adjacent supplementary teaching

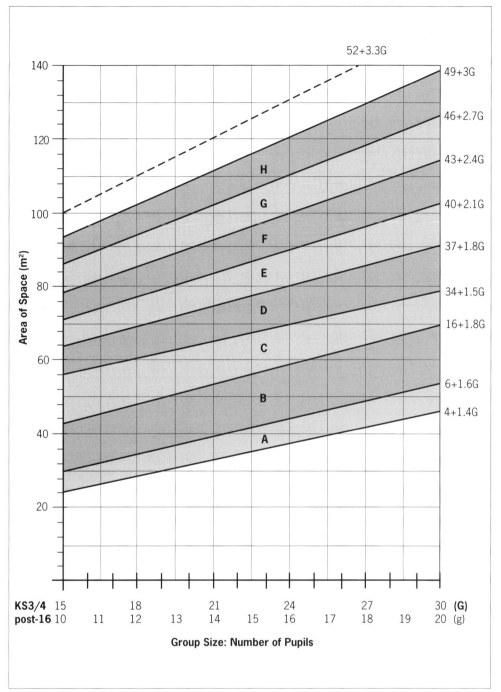

Figure 4.8: graph showing areas, zones A to H, for all general and practical spaces as follows:

standard classroom	A
large classroom	B
GNVQ/business room	C
IT room	C-D
science laboratory	D
food room*	G-H
multi-materials*	G-H
PECT*	F
textiles	E-F
graphics	E
2D art	D-E
3D art/textiles	F
music room	B
music recital room	D
drama/media studio	D-E
performance space	F+

The broken line above zone H indicates a further zone that may be appropriate in some unusual cases, such as where two practical activities are housed in one space, as described in paragraph 4.39.

In the formulae, G is the KS3 or KS4 group size. The sixth form group size (g) is shown on the lower line of the horizontal axis (g = G ÷ 1.5).

*group sizes are unlikely to exceed 25.

spaces such as a control room for stage lighting and sound systems, or a green room and dressing room facilities near to the stage area. Space will also be needed for storing equipment and costumes.

Physical Education Spaces[1]

4.59 All schools require an indoor PE space for at least one class group. Figure 4.9 illustrates the range of activities that are possible in sports spaces of different sizes. Approximately half of the curriculum time in PE will take place outdoors, particularly athletics and team games (see paragraph 5.4). Theory aspects of PE as an examination subject may be taught in a general teaching space. A PE space may have a dual use as the assembly or performance space, in which case the size of the space will also depend on the assembly group size.

4.60 A gymnasium space of 260m² will allow for a range of team games and gymnastics for a group of 30 pupils. If a second space is to be provided, this may

Note
1: The Handbook of Sports and Recreational Building Design: Volume 2 Indoor Sports (second edition), The Sports Council 1995, gives relevant guidance.

Figure 4.9: table showing possible activities in a range of PE spaces.

activity	badminton	basketball	hockey	volleyball	five-a-side	cricket nets	gymnastics	dance/movement	trampolining
size (m²)									
145	1 court							●	1 no.
240 - 260	2 courts	mini		●	●		●	●	2 no.
370 - 430	3 courts	●	●	●	●		●	●	4 no.
430 - 520	4 courts	●	●	●	●	4 no.	●	●	5 no.

be equipped for a different activity such as dance which will require a space of about 145m² for a group of 30 pupils.

4.61 Alternatively, a sports hall of 520m² can accommodate two groups of 30 for many activities and can also be used by one group for indoor sport requiring large 'pitches'.

4.62 As part of the PE facilities there may be supplementary facilities for small groups such as a fitness room. Approximately 1m² per workplace should be allowed for the storage of PE equipment.

Non-Timetabled Teaching Areas

4.63 Teaching spaces which are predominantly untimetabled include:

- halls used for assembly, examinations and public performances;
- the library resource centre and any local resource areas;
- various supplementary teaching spaces;
- social and study areas.

The number and size of these areas will depend on the philosophy and organisation of the school and some functions will overlap. These spaces must be considered within the teaching area available after all the timetabled spaces have been listed. Some schools may also have an area for individuals or small groups with special educational needs (paragraph 4.78).

Halls

4.64 Access to a large space such as a hall will be needed for assembly, examinations and music and drama performances to large audiences. Many of these activities will happen infrequently so it is best for some large spaces to be multi-purpose and perhaps partially timetabled. Some thought should be given to the position of these spaces in relation to the departments that may use them.

Assembly

4.65 All schools need to have space where pupils can gather for daily collective worship, in groups which may vary from a year group to, occasionally, the whole school. An area of 0.4m² to 0.55m² per pupil may be needed, depending on whether pupils are standing or sitting. Additionally, the area of any space used for assembly should allow for circulation and space at the front for small scale presentations. Unless it is partially timetabled, it is unlikely that a space with the ambiance of a traditional hall would be utilised sufficiently to be justified within reasonable gross area constraints if it is above 260m². In larger schools, spaces such as sports halls should be considered for occasional whole school gatherings.

4.66 An adjacent storage area is useful for chairs, rostra and similar equipment.

Resource Areas

4.67 The library resource centre is the main information base for the school. It can accommodate a wide range of activities including reading, IT use and group study. Whole class groups are usually taught induction courses for information-handling skills in the library. It is quite common to find the library partially timetabled for personal and social education (PSE) and for other subjects such as English or history.

4.68 Additional resource areas may be located within or close to departmental suites. Resources may include books and materials, accessible by pupils[1], related to the activities of the particular department. They often house clusters of IT equipment. These local resource areas are often combined with a supplementary teaching area where pupils can work individually or in groups away from the main class.

4.69 The graph in figure 4.10 shows a range that may be appropriate for the total area of all the school's resource spaces. This includes the library resource centre, which would normally occupy an area above the broken line on the graph, associated teaching spaces such as a careers space, and other resource areas around the school. This area would not include any timetabled spaces which are also used as bookable resource areas, such as IT rooms, or dedicated sixth form study areas, which may be adjacent to the library resource centre (see paragraph 4.80). The upper end of the area range could provide space for a library supported by local resource areas or an integrated library resource centre for an 11-18 school. In each case a careers area could be included in the total area. The lower end of the range may be more suitable for a school with no sixth form and no additional local resource areas.

The Library Resource Centre

4.70 Ideally the library resource centre should be able to cater for 10% of the total number on roll, engaged in a variety

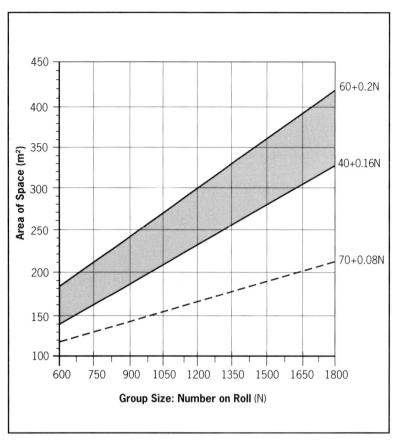

of activities at any one time. The area should be large enough or have a flexible furniture layout to allow whole class teaching when necessary.

4.71 To cater for a broad range of activities, the library resource centre may need space for:

- fiction and non-fiction books;

- a reference section including newspapers, periodicals (perhaps on CD-ROM) and resources such as charts, maps and photographs;

- IT work stations including CD-ROM (some perhaps in an adjacent space);

- multi-media kits or inter-active videos or IT;

- film and audio materials;

- artefacts, including models and specimens;

- informal seating;

- quiet study;

- study space for individual pupils or small groups;

- a control desk;

Figure 4.10: graph showing a range for the total area of the resource areas, including: the library resource centre (usually taking up at least the area indicated by the broken line); local resource areas in departments; any careers space. This area would not include any timetabled spaces also used as resource areas or sixth form study areas.

Note
1: Spaces used for resources that are predominantly accessed by staff count as non-teaching area.

Library Resource Centre Example (figure 4.11):
The plan below shows a library resource centre suitable for an 11-18 school with 900 KS3 and 4 pupils and about 200 sixth form pupils (as in appendix 3). It can accommodate a reasonable ratio of books to pupils of around 7.5:1, with some books on display, on the basis that a further 2.5 books per pupil are elsewhere. It has reading and reference areas, IT and audio-visual facilities and a display space. There is sufficient area for a whole class group to use the central facilities without disturbing others, including sixth form students, who may be working unsupervised or using resources elsewhere in the library. A smaller library may have a lower provision of all of these facilities, with fewer computers, and would be suitable for an 11-16 school with a similar number on roll.

Figure 4.11: an example of a library resource centre layout.

- a careers library (in or adjacent to the main space);
- display/exhibition area.

4.72 The overall layout should allow for a welcoming atmosphere, with natural light but sufficient wall space for full height shelving. It should allow general supervision from the control counter, which should be near to the entrance and may be near the office or workroom. One main entrance may be best for security. Storage for coats and bags also needs to be considered.

4.73 The following non-teaching spaces may be required to support the library resource centre:

- librarian's office;
- workroom;

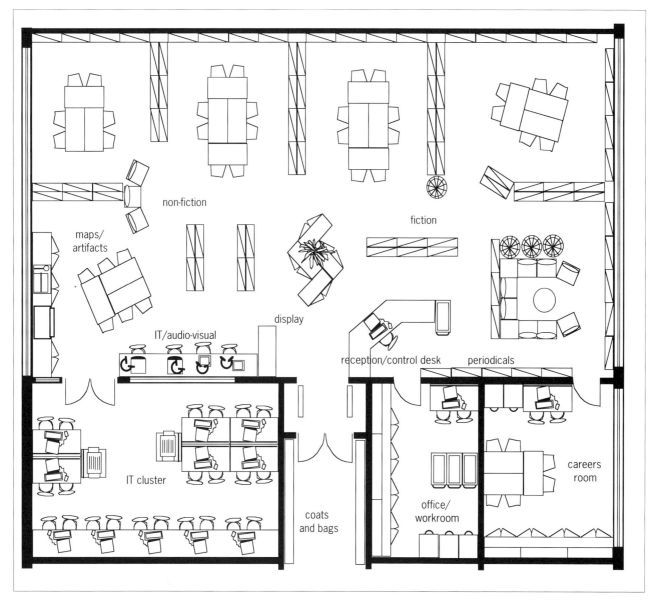

non-fiction

maps/artifacts

fiction

display

IT/audio-visual

reception/control desk

periodicals

IT cluster

coats and bags

office/workroom

careers room

- equipment and stock storage.

The librarian's office may be combined with the workroom, where books are sorted and catalogued. The storage area may be used for audio-visual aids as well as books in which case adequate security measures must be taken. A photocopier may be in the office/workroom or in the main space near the counter.

Local Resource Areas

4.74 The size of a local resource area depends on the range of activities being accommodated. An area in **zone A** (figure 4.8) accommodates a small cluster of computers in an area with minimal circulation. If the space also serves as a supplementary teaching area, such as a design/resource area (as in paragraph 4.76), an area in **zone B** may be needed. It is often economical to use resource areas for additional functions such as social bases or display, or to share them with partly timetabled spaces.

4.75 The storage needs of local resource areas are part of the overall teaching storage of relevant departments summarised in paragraphs 4.93 to 4.100.

Supplementary Teaching Areas

4.76 There is a range of untimetabled supplementary areas used by small groups for different activities. Spaces range from a group room for a foreign language assistant (FLA) to a dark room and have already been described under each subject in this section. These spaces may also be used as local resource areas. A design/resource area in a design and technology suite, for example, may provide display space alongside resources and a shared area for design.

4.77 The table in figure 4.12 shows the possible relationship between timetabled and untimetabled spaces. It demonstrates that more than one activity may benefit from access to the same type of supplementary space and that in some cases sharing could be considered.

supplementary areas	general teaching	science	textiles	multi-materials	food	PECT	business studies	art and design	music
IT/ resource/ display	○	○	○	○	○	○		○	
seminar room	○	○			○		○		
darkroom		○	○					○	
kiln room								○	
sixth form project room		○	○	○		○		○	
external project area		○			○			○	
food testing area					○				
microbiology room		○							
greenhouse		○							
animals room		○							
recording control room	○								○
music group rooms									○
FLA group room	○								○

Figure 4.12: table showing the possible range and location of supplementary teaching areas for practical subjects.

SEN Support Spaces[1]

4.78 Most schools have teaching support for pupils with special educational needs. This is usually in the form of assistance in timetabled lessons, but some individual or small group tuition may be done, often requiring a resource space. The number and size of rooms required depends on the particular needs of the pupils, but the area is likely to be in **zone B**.

Study and Social Areas

4.79 Some schools may choose to use a portion of the untimetabled teaching area for social areas such as 'house bases'. It is economic to partially share other timetabled rooms or link exhibition or circulation space to effectively create these spaces.

4.80 Schools with sixth forms usually provide an area for private or self-study for post-16 pupils. It may be adjacent to or in the same space as a common room, for more social activities, or deliberately separate. The demand for self-study area can be derived from the timetabled pupil time outside taught lesson time, assuming that sixth formers will use the library resource centre during lessons as well as in their study time (see appendix 3).

Note
1: See Building Bulletin 61: Designing for children with special educational needs: ordinary schools, HMSO 1984, now out of print but available from HMSO Books (Photocopies Section).

Non-Teaching Areas

4.81 There is a range of non-teaching facilities that schools might need in addition to the teaching storage described earlier in this section. Changes in priorities and school management have highlighted the importance of flexibility and adaptability in planning for these areas.

4.82 There can be a wide variation in the extent of non-teaching accommodation in existing schools and in briefs for new schools. Non-teaching spaces are important for the effective operation of the school, but they can easily consume too much of the total area. In section 1 it is suggested that between 40% and 43% of the total gross area might be devoted to non-teaching accommodation.

4.83 The table in figure 4.13 indicates an approximate percentage breakdown of the total non-teaching area based on groupings according to function. A percentage range is given which reflects differences of school organisation and numbers of pupils on roll. Pupils with special needs will also affect the provision. The final schedule of non-teaching spaces may need to be adjusted to suit the available gross area.

Figure 4.13: proportional breakdown of non-teaching areas.

type of space	proportion of total non-teaching area	proportion of gross area
staff and admin. accommodation	13% - 14%	5.2% - 5.6%
pupils' storage/washrooms	10% - 13%	4.0% - 5.2%
teaching storage	13% - 15%	5.2% - 6.0%
catering facilities	10% - 15%	4.0% - 6.0%
ancillary spaces	3% - 6%	1.2% - 2.4%
circulation and partitions	44% - 48%	17.6% - 19.2%
total non-teaching area		**40%**

Staff Accommodation

4.84 The range of accommodation provided for staff must include a head teacher's office, social and work area for teaching staff and staff lavatories. It might also include:

- deputy head teachers' offices;
- general office and secretarial rooms;
- heads of departments' offices or departmental bases and stores;
- senior administrators' or bursar's rooms;
- reprographics facilities;
- staff showers[1];
- central stockroom storage;
- staff changing facilities;
- medical or rest room.

All of these functions might be expected to occupy 13% to 14% of the total non-teaching area.

4.85 Head teachers and senior management will require space for meetings and interviews which may need to be confidential. Sometimes, records, valuables and cash are stored in these offices, in which case additional security may be necessary. There may be offices for faculty heads, possibly shared in small schools. The librarian may also have an office or workroom (see paragraph 4.73).

4.86 General office accommodation and secretarial facilities have to balance openness and accessibility for staff, pupils, parents and other visitors with the occasional need for security and confidentiality.

4.87 Staff social and work spaces provide an area for large group meetings, refreshment and relaxation, and for preparation, assessment and recording work. Care must be taken not to over-provide for these important but intermittent functions. In general, a staff room capable of holding 75% of full time staff for social facilities and 15% to 20% of full time staff for work space should be sufficient. This maximises the use of the spaces, and assumes that large meetings might take place in the teaching accommodation. Each departmental suite can benefit from a departmental base where staff prepare work, and where reference and audio-visual material is stored. This space may overlap with supplementary teaching areas.

4.88 Reprographic facilities can generate large demands for space, as the technology enables schools to produce most of their own resources. It might be sensible to contract out the reprographic work to specialist centres, with savings on staff costs, time and space consuming equipment.

4.89 Staff lavatory provision[1] should be separate from those for pupils. Facilities for the disabled may be unisex and used by adults and pupils (if it is in a self-contained room off a circulation space).

Pupils' Storage and Washrooms

4.90 These functions might occupy between 10% and 13% of the non-teaching area and include:

- pupils' lavatories;
- storage or lockers for personal belongings;
- areas for changing and showering;
- facilities for the disabled.

The statutory requirements for *pupils' lavatory provision* are set out in the School Premises Regulations; one sanitary fitting must be provided for every 20 pupils. In washrooms for boys, up to a third of these may be urinals. The proportion of wash-hand basins to sanitary fittings should be 1:1 in washrooms with two fittings or less, and at least 2:3 in washrooms with more than two fittings.

4.91 An area is needed for *pupils' storage or lockers* sufficient to allow for storing and drying of pupils' outdoor clothing and for storing baggage and personal belongings[2]. A combination of clusters of lockers and classroom coat and bag parks is generally more effective than central cloakrooms.

4.92 Pupils' changing and showering facilities must be provided. The area will

subject	area per workplace
• general teaching	0.08 - 0.09m²
• GNVQ/business	0.3 - 0.4m²
• IT	0.3 - 0.4m²
• science	0.4 - 0.5m² inclusive of preparation
• food	0.5 - 0.6m²
• multi-materials	0.6 - 0.8m² exclusive of preparation
• art and textiles	0.4 - 0.5m²
• music	0.3 - 0.4m²
• PE	0.09 - 1.00m²

Figure 4.14: table showing storage requirements for timetabled subjects.

relate to the amount of both indoor and outdoor sports facilities that can be timetabled for use at any one time. Normally, it is sufficient to provide a changing room for half a year group. Equal and separate facilities are needed for boys and girls. There is a growing movement away from communal showers in schools, particularly where privacy is required for religious reasons. The Sports Council recommends an area of 0.75m² per pupil changing and a ratio of one shower (occupying 1.7m²) to every seven pupils changing, or more for outdoor sports[3].

Teaching Storage

4.93 Storage additional to the local storage included in the area of teaching spaces (figure 4.8) is needed, either within the classroom or in separate walk-in stores. The area ranges in figure 4.14 can be used as a guide to storage requirements.

4.94 The higher end of the range may be suitable in smaller schools where a core of equipment and materials may be required. The lower figure may be appropriate where storage or preparation is concentrated into one space, for example in science.

4.95 There are more storage spaces in practical areas than in general teaching because of the range and quantity of

Notes
1: Workplace, Health Safety and Welfare Regulations 1992, Approved Code of Practice gives guidance on the minimum provision of sanitary conveniences for employees.
2: See A&B Paper number 15, Lockers and Secure Storage, DES 1990 and building bulletin 58: Storage of Pupils' Personal Belongings, HMSO 1980.
3: Handbook of Sports and Recreational Building Design: Volume 2 Indoor Sports (second edition), the Sports Council 1995.

material to be stored. This can include raw materials, equipment, work in progress, finished work and musical instruments. It may be necessary to control levels of humidity and temperature within some stores.

4.96 In science, preparation areas are needed for cleaning and sorting equipment. There should also be a materials preparation area associated with multi-materials activities, in an area additional to that shown in figure 4.14.

4.97 The area needs for drama storage will vary depending on the size of the school and the drama courses and extra-curricular activities offered. The need for green rooms and storage space for pupils' belongings may also be considered.

4.98 In addition to the storage needed for timetabled areas, there will also be general storage requirements for untimetabled spaces. Storage for chairs and rostra may be needed adjacent to the main hall.

4.99 Equipment and stock storage spaces may be required in the library resource centre. The storage needs of local resource areas are taken to be part of the overall teaching storage.

4.100 All of the teaching storage might be expected to occupy between 13% and 15% of the total non-teaching area.

Catering Facilities

4.101 The pattern of provision of school meals has changed considerably over the years. Schools are no longer obliged to provide cooked meals. This, together with further developments in food preparation, refrigeration and storage, have caused the reduction in size of school kitchens. Many are now 'finishing kitchens', simply cooking or heating convenience food. The scale of operation, number of staff employed and ancillary facilities such as a lavatory, office and rest room will vary greatly. Combined with the dining area the allowance within the total non-teaching area can vary between 10% and 15%.

4.102 The dining area is likely to be heavily used at lunchtimes and possibly at other break times. Allowing time for clearing up after meals, there may be scope for the space to be used for other teaching or non-teaching purposes and this could be allowed for in the timetable analysis (section 2). Current take-up rates suggest that, at most, 80% of pupils have school lunch. Generally, at least three sittings spread over the lunch break are possible, and an area of 0.9m² per pupil at any one sitting is a reasonable rule-of-thumb. In addition, there must also be sufficient area given to non-diners to eat their sandwiches. The scope to vary the overall catering provision lies within the priorities given to dining by the school and the management and organisation of the facilities.

Ancillary Spaces

4.103 Ancillary spaces include: the caretaker's office; maintenance storage; boiler or plant rooms; and cleaners' stores. The need for these spaces is dependent upon the management and organisation of the school. They might occupy between 3% and 6% of the non-teaching area.

Circulation and Partitions

4.104 The amount of internal partitions often relates to the level of circulation, and the design of the building has a great impact on the proportion of the area taken up by circulation. Wherever possible, designers should try to minimise the circulation areas. They cannot normally be used for any educational or social purpose and they simply add to capital and recurrent costs. However, the circulation routes need to be wide enough for fire safety and for pupils to move easily throughout the school between lessons. Corridors that are too narrow can lead to jostling and bad behaviour. The proportion of area that is occupied by circulation and partitions tends to operate inversely to the size of school and might range between 44% and 48% of the non-teaching area.

Section 5: School Grounds

This section outlines the main issues in the choice and design of school sites, and then gives guidance on areas. The overall **site area** comprises two types of predominantly **timetabled areas** (playing fields and hard surfaced games courts), **non-timetabled areas** (for informal and social use and for habitats) and the area taken up by the **buildings and access**.

5.1 The layout of the site should allow for various overlaps in the usage of these areas. For instance, hard surfaced games courts and the spaces around them may also be used for informal and social use. Figure 5.1 shows the type of environment that might be seen in each area and the opportunities that they can offer.

Figure 5.1: table showing some types of environments in school grounds and their possible use.

Type of Environment	time-tabled	non-timetabled									other areas	
	formal curriculum — physical education	science/environment	geography	English	mathematics	design and technology	humanities	art	informal curriculum	hidden curriculum	services only	
playing fields												
grass pitches/play area	●								●	●		
all-weather pitches	●											
hard surfaced games courts												
games courts	●								●			
multi-games/skills practice	●								●			
informal and social area												
soft: grassed areas	●	●						●	●	●		
landscaped areas/mazes		●	●	●	●	●	●	●	●	●		
hard: social areas at edges .				●					●	●		
marked street games area	●		●		●	●			●	●		
time-line/maps		●	●		●		●			●		
seating/shelter				●					●	●		
amphitheatre				●			●	●	●	●		
structures/sculpture		●		●		●		●	●	●		
habitat areas												
gardens/trees		●	●	●	●	●	●	●	●	●		
rock garden/vegetable plots		●		●	●	●	●	●	●	●		
wildlife habitats		●	●	●			●	●		●		
livestock enclosures		●		●	●	●	●	●		●		
nature trail		●		●			●	●	●	●		
orienteering trail	●		●							●		
ponds/streams		●	●	●		●		●		●		
weather station		●	●									
buildings and access												
soft landscape										●		
roads/pedestrian access										●	●	
car parking/service yards											●	
outdoor storage/bicycle sheds										●	●	

Opportunities

5.2 A school's grounds can offer many opportunities:

- for pupils' education, recreation and social development;
- for community use;
- for developing links with sports clubs.

The design and layout of the facilities should aim to meet the demands of the *formal curriculum*, including physical education (PE) and other National Curriculum subjects, as well as those of the *informal curriculum*, before and after school and during break and lunch times. The nature of the grounds can also influence the *'hidden curriculum'*; the messages and meanings which pupils and visitors 'read' from the ways in which a school's grounds are designed, used and managed.

Timetabled Areas

5.3 PE involves many activities, including games, athletics and outdoor and adventurous activities, which place demands on the grounds. It is usually the only subject in the formal curriculum which requires outdoor facilities to be timetabled.

5.4 The amount and type of provision for outdoor PE can be derived from the requirements of the formal curriculum in the same way as the curriculum analysis described in section 2 can derive the number and type of indoor spaces required. The examples discussed in section 2 and in the appendices assume that 50% of PE teaching will be timetabled outdoors. The number and type of facilities will also depend on various factors including:

- the statutory requirements for playing fields;
- the characteristics of the site;
- the indoor facilities available (including a swimming pool);
- the extent of extra-curricular and community use.

Type of pitch or court		Age Range	length (m)	width (m)	end margin (m)	side margin (m)	comments
Association Football							
mini-football	5-a-side	8 - 14+	30 - 40	18.5 - 28	-	-	margins not applicable as play is usually to rebound walls
football	small	8 - 13	70 - 80	40 - 50	6.0	4.5	
	medium	13 - 15	75 - 82	45 - 55	6.0	4.5	
	large	15 - 16+	91 - 96	50 - 59	6.0	4.5	
	medium	18 and over	96 - 100	60 - 64	9.0	6.0	
	adult	18 and over	100 - 110	64 - 75	9.0	6.0	recommendation of FA UK
Basketball							
basketball		12 and over	28	15	1.0	1.0	under 12 court for mini-basketball the same, with double sized teams
Hockey							
mini-hockey	6-a-side	10 - 12	36 min	18 - 22	-	-	6-a-side is for recreation (mainly indoors) with no margins as play is to rebound walls.
	7-a-side	under 14	50.3 - 54.9	36.5 - 45.7	4.57	3.0	
hockey	small	11 - 13	73.2	45.7	4.55	2.75	
	medium	13 - 15	82.3	50.3	4.55	2.75	
	large	15 - 16	91.4	50.3	4.55	2.75	
	club	16 and over	91.4	55.0	4.57	3.0	
Netball							
first-step netball		8 - 11	15.25	10.15	0.75 - 2.0	0.75 - 2.0	first step court dimensions may be scaled down proportionately
netball	full-size	12 and over	30.5	15.25	0.75 - 2.0	0.75 - 2.0	
Rugby League Football							
mini-league		under 9	60	40	5.0	2.0 - 5.0	) in-goal dimension not fixed as
mod-league		under 11	50 - 80	35 - 60	5.0	2.0 - 5.0	) goals are set outside play area
rugby league	small	11 - 13	60 - 80	35 - 50	5.0	2.0 - 5.0	in-goal dimension 6 - 8m
	medium	13 - 15	70 - 90	40 - 60	5.0	2.0 - 5.0	in-goal dimension 6 - 8m
	large (adult)	15 and over	80 - 100	55 - 68	6.0	2.0 - 6.0	in-goal dimension 6 - 11m
Rugby Union Football							
mini-rugby		under7 and under 8	30	20	3.0	4.5	in-goal dimension 2m
		under 9 and under 10	59	35	3.0	4.5	in-goal dimension 5m
		under 11	59	38	3.0	4.5	in-goal dimension 5m
		under 12	59	43	3.0	4.5	in-goal dimension 5m
rugby	small	11 - 13	75	46	3.0	4.5	in-goal dimension 6.5m
	medium	13 - 15	82	50	3.0	4.5	in-goal dimension 6.5m
	large	15 - 16+	91 - 96	55 - 59	3.0	4.5	in-goal dimension 9.0m
	adult	16 and over	100	69	3.0	4.5	in-goal dimension 10 - 22m
Tennis							
short tennis		5 - 11	13.4	6.1	1.8 - 2.6	1.5 - 2.1	
padder tennis		5 - 11	11.9	5.5	1.8 - 2.6	1.5 - 2.1	four games accommodated on one full size tennis court
Lawn Tennis Assoc. court		11 - 16+	34.75	17.07	(5.49)	(3.05)	margins used in play, so margin dimensions included in court size
Volleyball							
mini-volley ball		9 - 13	9	6	3.0	3.0	can be played within one third of full-size court
volleyball		11 and over	18	9	3.0	3.0	preferred surface grass or sand
Multi-Games Areas							
junior	enlarged netball	8 - 11	39.5	27.5	-	-	dimensions include margins (space for other games)
senior	3 netball/3 tennis	11 - 16+	55	35	-	-	dimensions include margins
	3 netball/4 tennis	11 - 16+	65	40	-	-	dimensions include margins

				radius of outfield		
Cricket						
pitch size determined by age	under 7	14.63	)			cricket squares and artificial
	under 9	15.54	)	30m or to fit available space		wickets are normally located
	under 10	16.46	)			between winter games pitches
	under 11	17.37	)			
	under 12	18.29	)	35 - 37m (junior NCC)		
	under 13	19.20	)			
	under14 and under 15	20.12		40 - 46m (senior NCC)		

Playing Fields

5.5 Playing fields contain summer and winter games pitches and other provision such as cricket nets and athletic facilities, as listed in figure 5.2. For effective land use and safety, pitch layouts should include adequate margins around the playing area, especially where pitches adjoin, as recommended by the relevant national games governing bodies. Margins include run-off areas around pitches and space to allow realignment of pitches to compensate for wear. Larger clearances may be needed where pitches are close to roads or buildings, or to minimise the likelihood of balls falling onto adjacent land. The broken lines on the graphs in figures 5.8 and 5.9 indicate the statutory requirements for playing fields, which include margins.

5.6 There will also normally be other areas around the perimeter of the playing fields that are unusable for games due to their shape or other characteristics and not therefore counted against statutory requirements. Such spaces can be developed as habitats, as described in paragraph 5.21, or for informal and social use.

5.7 There is no statutory requirement for playing fields for pupils under the age of eight, but infant schools may benefit from some grassed space for skills development and small games, within an area similar to that of hard surfaced games courts in junior schools (as shown in figure 5.3). For primary schools, the playing field area will include one or more marked pitches. A straight running track with six or eight lanes and a length of 60 to 80m is useful for summer use. At secondary level, the area should provide winter pitches for team games chosen by the school. In summer, cricket pitches and a 400m running track may be laid out. The track should avoid the heavily used areas of the winter pitches, such as goal-mouths.

5.8 A statutory requirement is for playing fields to be of a quality to sustain at least seven hours of use per week in term time. This will require an adequate specification, including good drainage. Heavier use may require an improved pitch specification.

Hard Surfaced Games Courts

5.9 Hard surfaced areas for games courts, and any skills practice areas adjacent or overlapping, should be level, drain well and have an even surface which is free of obstructions. Tarmacadam is usually used, but other surfaces (as in paragraph 5.12) are also possible. Areas should be of a shape and size suitable to allow courts to be marked out, with reasonable margins, to the dimensions set out in figure 5.2. The graph in figure 5.3 shows a range of areas for hard surfaced games courts for all types of schools. However, infants are unlikely to need formal courts, as the hard surfaced portion of the area for informal and social use can be laid out for skills learning and small games practice.

Multi-games areas

5.10 For juniors, a court for organised games is valuable, such as a full-size netball court. Where space is limited, netball can be played on a slightly smaller court by scaling down the dimensions. The 'grids' formed by the markings can also be used for mini-games such as short tennis. Any other hard surfaced areas used for PE need to be close to the court to enable the teacher to maintain effective supervision.

Figure 5.2 (opposite): recommended dimensions of courts and pitches used in schools, related to age. Margin dimensions should be added, in all cases except lawn tennis courts, to both ends and both sides.

Figure 5.3: hard surfaced games courts: graph showing range of areas for the hard surfaced games court or courts for all types of schools, except nursery, infant and first schools which are unlikely to require formal courts.

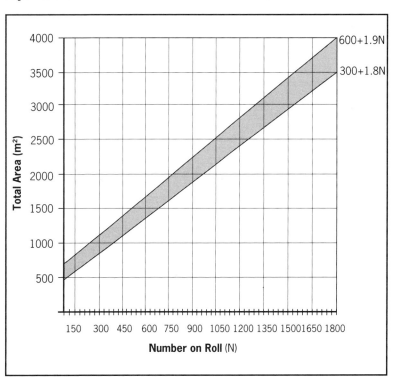

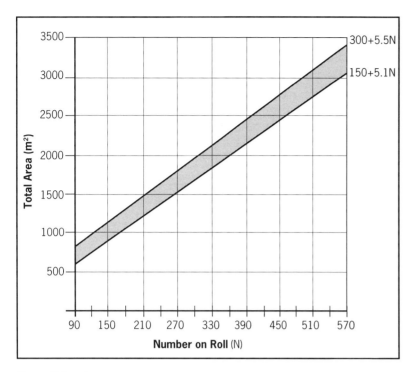

Figure 5.4: primary informal and social areas: graph showing range of areas for informal and social use for primary and middle schools. Individual judgements may be made about schools with fewer than 90 pupils.

5.11 At secondary level, the hard surfaced provision will often include a range of multi-games courts. By laying out a variety of courts within a single area, supervision is made easier and the range of games extended. For example, three or four tennis courts, overlaid on to netball courts, can also be used for mini-hockey or five-a-side football.

All-weather pitches

5.12 A range of artificial surfaces is available for games courts and pitches that can be used more intensively and immediately following rain. The surfaces may be hard porous (water-bound), not commonly provided now as maintenance needs can be high, synthetic turf or polymeric[1].

5.13 The choice of surface should be based on performance, safety and durability, through:

- the properties best suited to the types of games to be played, such as the 'ball bounce';
- slip resistance and abrasiveness;
- wear resistance;
- maintenance requirements.

The location of all-weather pitches should allow pupils to avoid treading in mud by having to walk over grass to reach them.

Non-Timetabled Areas

5.14 Non-timetabled areas include spaces for informal recreation and social use and for natural habitats. They can offer opportunities for the formal and informal curriculum (summarised in figure 5.1).

The formal curriculum

5.15 Apart from PE, many subjects in the formal curriculum can benefit from the use of a variety of outdoor spaces and habitats, usually on an irregular basis. For instance, there are possibilities for the teaching of skills such as:

- observation and classification in science;
- mapping and recording in geography;
- communication and comparison in English and humanities;
- estimation and analysis in mathematics;
- surveying and design in design and technology or art.

5.16 Part of the formal curriculum may often be served through the provision of outdoor seating or an amphitheatre. A time-line or map marked on hard play areas can be used in mathematics, humanities and geography, while structures or sculpture can enhance art. A pond may provide a resource for teaching about life processes and living things in science or about contrasting environments in geography. Plants, animals and soils can be investigated thoroughly if pupils have direct experience of natural habitats.

The informal curriculum

5.17 Informal learning and recreation also take place before and after school, and during break and lunch-times. This may occupy a quarter of a pupil's school day and much of this time will be spent outside. The character of a school's grounds can affect pupils' behaviour. Grounds which provide diversity of provision, including seating and sheltered areas, can allow pupils the maximum opportunity for creative play and social development.

Informal and Social Areas[2]

5.18 A significant portion of the site should offer a variety of hard and soft areas to suit the formal curriculum and the informal and social activities of pupils during breaks and before and after school. These areas should be conveniently situated and robust, but should also provide shade and shelter. They might include site furniture and a suitable landscape, including smaller, more intimate areas. Ranges of areas for informal and social use are shown in the graphs in figures 5.4, for primary and middle schools, and 5.5, for secondary schools.

5.19 Approximately half of this space should be hard surfaced area additional to that for games courts. In primary schools, this might be marked out for street games and geometric designs and infant pupils may use this for PE (see paragraph 5.9). In secondary schools, it should be large enough for boisterous play, while quieter, sheltered bays around the edge enable separate activities to take place without isolation. The remaining half of this area is likely to offer grassed space and soft landscaping suitable for pupils' use (which may overlap with habitat areas). It might also include an amphitheatre or sculpture.

5.20 Spaces around the school building can often serve as social areas if site furniture is provided. A position near to the building is generally desirable for ease of access and supervision, although care has to be exercised over the proximity of ball games to windows. In primary schools, a patio close to the building will often act principally as a social or teaching area, while the active play areas are beyond it. Groups of seats in circles or squares facing inwards, or low walls, can increase the opportunities for pupils to sit down and socialise.

Habitat Areas

5.21 There is benefit in all sizes and types of school for a proportion of the grounds to be developed for a wide range of activities, including wildlife habitats, garden areas, livestock enclosures, and

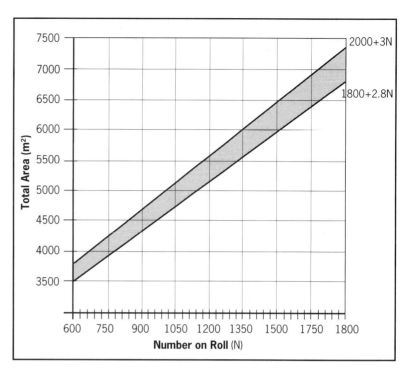

natural landscapes. Equally, different habitats, for example, an historical garden or an orienteering trail, can support particular subjects. They can also enhance play or recreational time. Increasingly, habitat areas are being given a central and accessible location, rather than being seen as optional extras. However, some wildlife areas should normally be undisturbed, so are best positioned away from busy social areas. Areas that are unusable for playing fields due to the shape of the site may be suitable for habitat areas.

Other Areas

5.22 Remaining areas will include the footprint of the buildings and areas that are generally unsuitable for pupils' use, except for access. These are all grouped under the following heading.

Buildings and Access

5.23 This includes the footprint of the school buildings, access roads, service yards, car parks, paths and outdoor storage. Some soft landscape may also fall within this area if it is not suitable for other use. The proportion of the total site occupied by this area is likely to be higher in a small school than in a large one of the same type.

Figure 5.5: secondary informal and social areas: graph showing range of areas for informal and social use for secondary schools. Individual judgements may be made about secondary schools with fewer than 600 pupils.

Notes

1: **Hard porous (water-bound):** a finely crushed surface of clay bound material, stone or blaes, on a base of coarse clinker, ash or crushed stone. Under pitch drainage is usually provided.
Synthetic turf: coarse plastic pile, usually sand-filled, on a plastic or rubber shockpad backing, laid on a prepared base of concrete, macadam or unbound stone. Under pitch drainage is normally provided.
Polymeric surfaces: shredded rubber (most commonly), wood fibres or granules of cork, rubber and plastics bonded with bitumen, latex or polyurethane, on a base of concrete or macadam laid to fall.
2: The area for informal and social use added to either the area of hard surfaced games courts (for schools with pupils over the age of eight) or the same area of grassed space (for infant schools), plus some habitat area, is generally equivalent to the 'recreation area' defined in the previous Education (School Premises) Regulations (1981).

Layout of the Site

5.24 Planning the layout of any school site should involve the consideration of a wide range of options before one is chosen for further development. The landscape design should provide a framework, yet allow for the school to develop its grounds gradually, with the participation of pupils. The playing fields and the area of buildings and access are two major determinants of the size and characteristics of the remainder of the site.

5.25 The relationship between indoors and outdoors is most important. The location, configuration, layout and servicing of the school buildings fundamentally affect the creation and effective use of external space. When the indoor and outdoor spaces are planned together, the effect that different building locations and orientations have on the design of the remainder of the site should be appraised before a final decision is reached. If later extension of the building is envisaged, it is better to identify from the start where this could be sited.

Figure 5.6: example of site layout of a 315 place primary school.

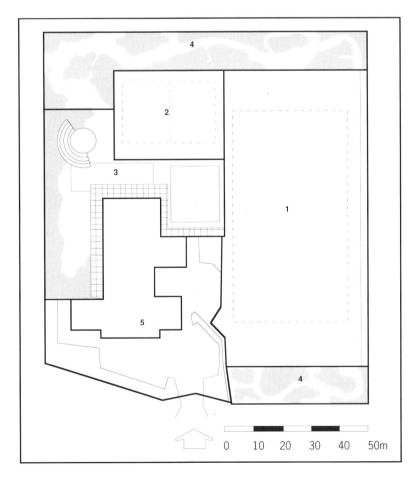

5.26 The site layout should take account of:

- the nature of the site, including its shape, contours, subsoil and exposure to wind, rain and noise;

- the need for supervision, often over more than one activity at a time;

- trespass, vandalism and security;

A Primary School Example (figure 5.6):

This example (left) illustrates a 5 to 11 primary school for 315 pupils. The total site area of 13,800m² is in the middle of the area guidance in figure 5.8, between 12,922m² (3000 + 31.5 x 315) and 14,725m² (3700 + 35 x 315). The statutory minimum playing field area for the 135 pupils over eight years of age is 5,000m². Separate hard games and grassed provision are provided for infants and juniors.

1. Playing Fields

5000m²   36% of total site area

A 70m x 40m grass pitch is provided for winter games by junior pupils. In summer, an 80m straight running track is marked out, with other grass areas used for cricket practice and games of rounders.

2. Hard Surfaced Games Courts

1100m²   8% of total site area

Juniors use a multi-games area incorporating a variety of court markings within a netball court. This is also used for informal recreation.

3. Informal and Social Area

2000m²   14.5% of total site area

Infants use a multi-use playground marked out with street games and geometric designs for PE. With play features such as boats and stepping stones, it is also used by all pupils as informal and social area. They also use a grassed space and areas nearer the building, which provide soft landscaping and terraced seating which can also be used for demonstrations and drama.

4. Habitat Areas

2300m²   17% of total site area

Around the edge of the site, there is a nature trail through copsing, a wildflower meadow and an orchard, as well as experimental plots and a pond.

5. Buildings and Access

3400m²   24.5% of total site area

A single entrance leads to a car park and a compact service area. The pedestrian site entry is separate from vehicular access.

A Secondary School Example (figure 5.7):

This example illustrates a secondary school for 1000 pupils. The total site area of 73,500m² is in the middle of the area guidance in figure 5.9, between 70,000m² (14000 + 56 x 1000) and 76,000m² (16000 + 60 x 1000). The statutory minimum playing field area is 45,000m².

1. Playing Fields

46000m²   62.5% of total site area

Winter pitches (shown with broken lines) comprise three small, three medium and one large, taken as football pitches in the layout. Alternatively, rugby or hockey could be provided, although the pitch dimensions vary. Summer facilities (solid lines) include one cricket square and one non-turf cricket pitch, set between winter pitches. A 400m athletics track is marked out on the grass, avoiding rough areas such as goal-mouths where possible. Cricket nets, high jump and long jump provision are accommodated at the perimeter (not shown).

2. Hard Surfaced Games Courts

2300m²   3% of total site area

A multi-games area is provided, which can be used as three netball courts, three tennis courts or one five-a-side football pitch.

3. Informal and Social Area

4800m²   6.5% of total site area

Hard surfaced areas are provided for boisterous activities and an amphitheatre area can be used for break-time activities as well as drama or music. Nearer the buildings, paved areas accommodate seating adjacent to planting for socialising and for an outdoor classroom during summer. Partially planted soft areas are between the buildings and playing fields for educational and social purposes.

4. Habitat Areas

10200m²   14% of total site area

Defined habitat areas are provided for garden areas and woodland, including nut and willow copsing and orchards, as well as a nature trail, a pond, wetlands and wildflower meadowland. Incidental habitat areas are located on grass areas adjacent to the playing field unsuitable for pitches and in the perimeter hedges and trees, where wildlife habitats and an orienteering trial are located.

5. Buildings and Access

10200m²   14% of total site area

The buildings are close to the road and the single main access. Two parking areas are provided for the school and for community use of the sports facilities.

Figure 5.7: example of site layout of a 1000 place secondary school.

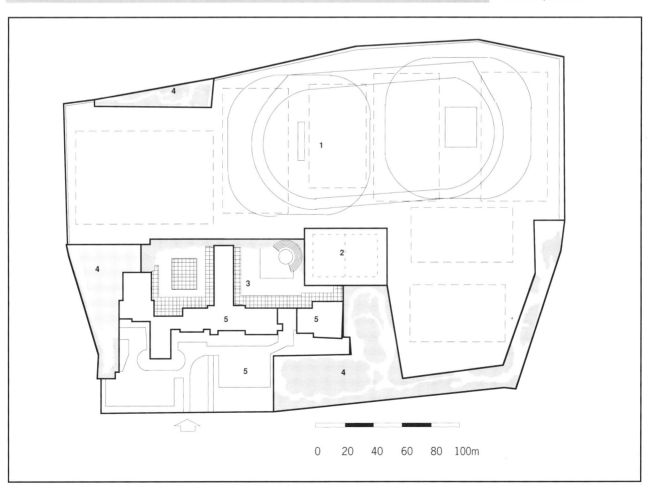

0   20   40   60   80   100m

- landscape quality and serviceability;
- access for the physically disabled;
- safety in terms of circulation, and location of activities;
- relationship between changing rooms and PE facilities;
- community use (see paragraph 5.29).

Pitch and Court Layouts

5.27 Careful attention should be given to the layout of pitches, courts and practice areas. Their location, size and shape should be based on a number of considerations, including the:

- statutory requirements for playing fields;
- safety considerations, including pitch margins and the direction of play (for example for cricket nets);
- gradient (a uniform fall of about 1:100 is ideal, but an even fall of up to 1:50 is possible, or more if it is across the line of play);
- layout of winter games pitches and their relationships to summer athletics and cricket provision;
- orientation of pitches (a north-south direction is desirable for most games);
- accessibility for maintenance equipment and water if irrigation is needed.

5.28 Two examples of typical site layouts are illustrated in figure 5.6, a 5-11 primary school, and figure 5.7, a secondary school. The associated text indicates the choices made in the design and the proportion of site available for each of the five types of area already outlined. In both cases, an economic approach to the layout design has confined the buildings and access to a compact area near to the site entrance and the shape and position of the building provides a protected environment for the immediate outdoor activities on hard paved areas. There is a progression from the building to areas for informal and social use, then to hard surfaced games courts, and finally to playing fields and habitat areas.

Extra-Curricular and Community Use

5.29 Many schools run activities at the end of the school day which may place demands on the grounds. The greatest potential for community use may well lie in the hard games courts or multi-games area (paragraph 5.10), especially where floodlighting is provided. Schools may have joint-use agreements with a local authority or with another organisation such as a local sports club. The key issues are the quality, durability and viability of the facilities. This means good access, including reception and parking, self-contained changing rooms and sufficient financial return to at least cover the marginal costs[1].

Site Area Formulae

5.30 The establishment of a target for overall site area, including playing fields, is important when considering a new school. It can also provide a valuable yardstick for considering improvements to existing schools. The graphs in figures 5.8 and 5.9 show ranges of total site areas for 5-11 primary schools and all types of secondary school, based on analysis of existing school sites and the likely needs of new school grounds. The ranges allow for variation in the shape and contours of the site and in the size of the building complex. They are based on formulae for each key stage (KS), like those in section 1, although in this case KS1 and KS2 have different formulae due to the statutory requirements for playing fields for pupils aged eight or over.

5.31 Area ranges for types of schools other than those used in the graphs can be derived by using the following formulae in the appropriate proportion:

Key Stage 1 total site area (m^2):

upper line	$3500 + 21N$
lower line	$2800 + 17.5N$

Key Stage 2 total site area (m^2):

upper line	$3850 + 45.5N$
lower line	$3150 + 42N$

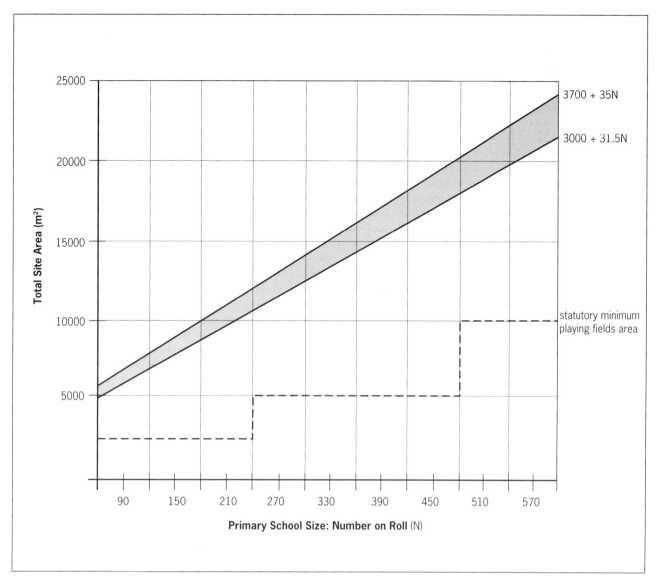

Primary School Size: Number on Roll (N)

3700 + 35N

3000 + 31.5N

statutory minimum playing fields area

Key Stage 3 total site area[2] (m^2):

upper line   8000 + 70N

lower line   6000 + 66N

As there is little difference in the demand for site area in KS3, KS4 and post-16 pupils, the formulae in figure 5.6 can be used for all types of secondary schools:

Key Stage 3/4 total site area (m^2):

upper line   16000 + 60N

lower line   14000 + 56N

In all cases, N is the total number on roll, including sixth form pupils and reception classes, which should be counted as KS1 for this purpose, but not nursery classes.

5.32 These general recommendations are not appropriate for primary schools with less than 60 pupils or secondary schools smaller than 450 (and similar pro-rata

figures for other types of school), for which individual judgements must be made. Appendix 4 gives formulae for a variety of types of schools, based on this method.

Nursery Pupils

5.33 Nursery schools and nursery classes are not included in the site area per pupil guidelines because of their widely differing needs and circumstances. The site area for a nursery class or a unit sharing the access and car park of an infant or primary school is likely to be at least 15m^2 per pupil. This includes about 9m^2 per pupil for an outdoor play area, which would be mainly hard surfaced. A free-standing nursery school may require 25 to 50m^2 per pupil or more depending on factors such as

Figure 5.8: primary school site area: total site area guidelines for 5-11 primary schools. The broken line indicates the statutory requirement for playing fields under the School Premises Regulations. Due to the relatively large incremental step in the statutory playing fields minimum at a NOR of 469, schools above this may need more site area than indicated.

Notes
1: Refer to 'Our School - Your School' DfEE 1996
2: The formulae for KS3 are used as a device to cover middle schools, which are generally smaller than the range of NOR in secondary schools, covered by the KS 3/4 formulae. They do not indicate that KS3 pupils have differing requirements from those in KS4.

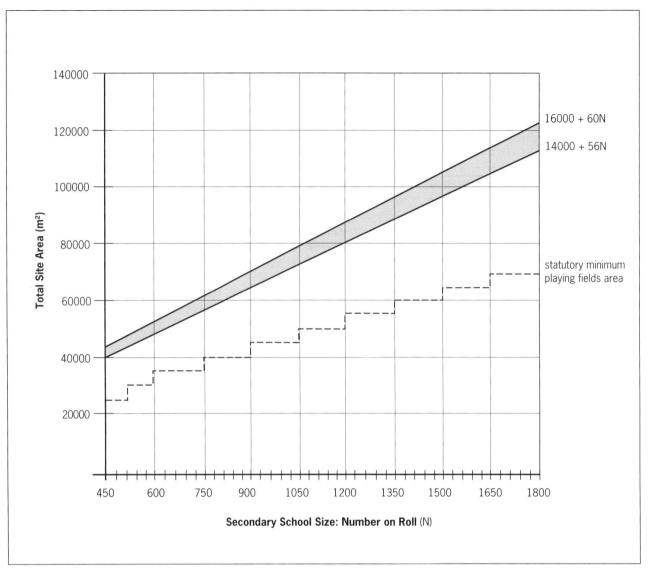

16000 + 60N

14000 + 56N

statutory minimum
playing fields area

Total Site Area (m²)

Secondary School Size: Number on Roll (N)

Figure 5.9: secondary school site area: total site area guidelines for all secondary schools (not including middle deemed secondary schools). The broken line indicates the statutory requirement for playing fields under the School Premises Regulations.

pupil numbers, the extent of car parking needed, and whether it is combined with day care, community or other facilities. In larger nursery schools, some allowance may be expected for economy of scale and the proportion of soft play area would increase. At the time of writing, more detailed guidance is being prepared.

Confined sites

5.34 Where available land is limited, the disadvantages of a restricted site may need to be weighed against the merits of a particular location. Where the site is below the recommended area range, the shortage of space can be offset to some extent by an increase in landscape quality to allow more intensive use. This is likely to mean more hard surfaced area, such as

a hard porous or synthetic surface that can be used for both PE and recreation.

5.35 In some situations, the playing fields may need to be off-site, perhaps some distance away. In this case, the guidance area figure minus the statutory playing field area would normally be appropriate for the main site. However, sufficient land may not be available to meet the statutory area requirements for playing fields. The School Premises Regulations allow hard porous pitches to be counted as double their actual area as they can support more intensive use. Similarly, synthetic pitches may allow the statutory area requirement to be reduced further. The statutory area may also be reduced where there is regular access to additional sports provision, such as a swimming pool, community sports centre or joint-use facility.

Appendix 1: A 5-11 Primary School Schedule

This appendix details a method for creating a schedule of accommodation for a primary school, using a typical example of a 5-11 school increasing in size to 14 classes.

6.1 First, the gross area and teaching area range is estimated using the guidance in section 1. Then a timetable analysis is used to identify the curriculum demand for any timetabled spaces, as outlined in section 2. These areas, together with those of the other teaching spaces and non-teaching spaces, are determined from the guidance in section 3 and are added to the schedule. Finally, adjustments are made to the overall schedule to ensure that it conforms to the original area constraints.

6.2 The example, based on a real school, was built for 315 infant and junior pupils. The method described below is used to develop a schedule of accommodation, as part of a design brief for a final number on roll of 420. The method described recognises the way that primary schools tend to be organised at present, but other methods, such as that in appendix 2, could also be used.

Overall Areas

6.3 Using the formulae in section 1, the range of gross area for a school of 420 can be seen to be 1598m² to 1796m² in total[1]. In a school of this size it is reasonable to expect considerable economies in the scale of provision of non-teaching areas such as offices and kitchen facilities. A teaching area exceeding 62% of the gross area could therefore be achieved. In this example 62% amounts to between 991m² and 1114m². This teaching area will be predominantly the basic teaching area as described in section 3 plus the timetabled teaching spaces and any untimetabled teaching areas.

Curriculum Analysis

6.4 In order to establish the demand for the timetabled spaces, a curriculum analysis is used. It follows the basic principles outlined in section 2. There are

two key differences: namely how the information necessary to put these principles into practice is obtained, and the assumption that the spaces required are related to the activities to be housed and not the subjects.

6.5 Many primary schools have non-teaching assistants (NTAs) who may supervise groups of pupils independently whilst remaining under the control of the teacher. It is therefore useful to calculate the space requirements for small groups withdrawn from the class in specialist or discrete spaces supervised by an NTA or teacher.

6.6 It is possible to calculate the number of rooms from the total supervising time required, as the equivalent of teacher periods as defined in paragraph 2.17. However, the percentage of curriculum time that is most likely to be known is the proportion of total pupil time[2], so the method used identifies the percentage of total pupil time spent in each type of space and then uses the average group size to determine the number of spaces required, as in paragraph 2.24.

Step 1: Distribution of Pupil Time

6.7 The first step of the analysis is to assess the average amount of time, in an average week, that each pupil needs to be supervised in each type of specialist or discrete space.

6.8 This may be known from an agreed brief, or may be derived from an analysis of a proposed timetable or programme of work. If the week is assumed to be divided into a number of roughly equal sessions, these can be used to determine the percentages of time that a typical pupil spends in the various types of teaching spaces. This may be derived by first analysing the average of the pupil time:

• for each year group; or

• for each predominant subject; or

• for different key stages, as appropriate.

Notes
1: The total gross area, based on the formulae in section 1, will range from
(420 x 3.4) + 170 =1598m²
to
(420 x 3.8) + 200 = 1796m²
2: This will not match the proportion of teacher time as the number of pupils being taught at any time will vary with subjects and age groups.

6.9 In this example, all pupils will either be in assembly or at break or in another non-teaching activity. It is not therefore necessary to include assembly in the pupil time assessed. If assembly is timetabled for some classes while others are in lessons, assembly should be regarded as a timetabled activity.

6.10 In the example used here, infants and juniors are taught for differing lengths of day, so these times have been calculated separately (in figures 6.1 and 6.2) and the numbers of spaces added together at the end. Some weeks may involve special events, such as preparations for Christmas. These unusual weeks can be ignored for the purposes of this exercise.

Average Group Size

6.11 In order to calculate the number of spaces from the curriculum percentages, the average group size for each space-type must be used. A reasonable average must be estimated, bearing in mind that the actual group size at any time may vary above or below this figure. The maximum likely group size (slightly higher than the average) will determine the size of space required.

6.12 In primary schools, a number of different activities are usually carried out at the same time by, for instance, four or five groups in a class. Most of these activities will take place in the basic teaching area, but one or two may be in a withdrawn group, supervised by an NTA or another teacher. The information required is: how often and for how long such sub-groups are to be provided with a separate supervised space, away from the rest of the class.

6.13 It is possible to measure the average group size as a number of pupils. However, it is often easier to use the class as the base unit and define the group size as a fraction of the class, as in figure 6.1 and 6.2. This also allows a broad range of group sizes to be encompassed within a simple average and avoids any complication caused by different sized classes. Hence, a group made up of a quarter of the class might be between six and nine pupils. In this case, average class sizes of 30 mean group sizes of seven to eight.

6.14 It should be noted how the group size affects the demand for spaces. For instance, if there are of about six pupils (a fifth of a class) in each group, the number of times the groups need to meet is five times more than the number would be if they met in a whole class group.

Figure 6.1: steps 1 and 2 for the infants department: (four classes rising to six) based on the average proportion of pupil time per week (excluding assembly) and an average group size across all three years.

Number of infant classes increasing from 4 to 6

SPACE TYPES	pupil time spent in space-type	average group size	number of spaces calculated for 4 classes	number of spaces calculated for 6 classes
	(as a percentage)	(as fraction of class)	$\frac{\% \times \text{no. of classes}}{\text{group size}}$	$\frac{\% \times \text{no. of classes}}{\text{group size}}$
class bases in basic teaching area	83.1%	1 (up to the whole class)	3.32	4.97
timetabled supplementary areas:				
mixed use practical	4.0%	0.25 (7-8 pupils)	0.64	0.96
food	1.5%	0.25 (7-8 pupils)	0.24	0.36
small group withdrawal	3.8%	0.25 (7-8 pupils)	0.61	0.91
studio	3.8%	1 (whole class)	0.15	0.23
hall	4.8%	1 (whole class)	0.19	0.29
outdoors	2.8%	1 (whole class)	(0.11)	(0.17)
TOTAL	100%		5.15	7.72

FTE number of teachers and NTAs in contact with pupils for 6 classes = 7.71

Number of junior classes increasing from 6 to 8

SPACE TYPES	pupil time spent in space-type (as a percentage	average group size (as fraction of class)	number of spaces calculated for 6 classes ($\frac{\% \times \text{no. of classes}}{\text{group size}}$)	number of spaces calculated for 8 classes ($\frac{\% \times \text{no. of classes}}{\text{group size}}$)
class bases in basic teaching area	79.75%	1 (up to the whole class)	4.78	6.38
timetabled supplementary areas:				
mixed use practical	4.00%	**0.25** (7-8 pupils)	0.96	1.28
food	1.25%	**0.25** (7-8 pupils)	0.3	0.4
small group withdrawal	3.75%	**0.2** (2-4 pupils)	1.12	1.5
studio	3.75%	**1** (whole class)	0.22	0.3
hall	5.00%	**1** (whole class)	0.3	0.4
outdoor	2.50%	**1** (whole class)	(0.2)	(0.2)
TOTAL	**100%**	-	**7.68**	**10.26**

FTE number of teachers and NTAs in contact with pupils for 8 classes = 10.5

Figure 6.2: steps 1 and 2 for the junior department: (six classes rising to eight) based on the average proportion of pupil time per week across all subjects and an average group size across all four year groups.

Step 2: Calculated Number of Spaces

6.15 Having established the information in step 1, the number of spaces demanded can be calculated in the next column by multiplying the percentage of pupil time by the number on roll and then dividing by the group size, as in paragraph 2.25.

6.16 The number of spaces relates to the number of separate pupil groups, supervised by a teacher or NTA, scheduled at any one time. Primary schools must bear in mind that small groups withdrawn to discrete space types will require supervision by teaching staff or NTAs[1]. It would clearly be a waste of area and resources to provide such spaces if the associated demand for supervising time was unrealistic. Average group sizes of less than five are likely to require an uneconomic amount of supervision time.

6.17 As a rough check, the total number of spaces calculated should be approximately equal to or below the full time equivalent (FTE) number of teachers and NTAs[2]. If not, the number of staff may have to be increased or the demand for rooms reduced by raising some average group sizes. As figures 6.1 and 6.2 show, the calculated number of spaces for the increased number of classes in the example (six infant and eight junior) is proportionally higher than that for the original numbers.

Step 3: Rounded Number of Spaces

6.18 At this point, the actual number and type of spaces required can be decided, based on a reasonable frequency of use, by rounding up the calculated figures.

6.19 For the basic teaching areas, the number of class bases will almost certainly need to be the same as the number of classes in the school. The calculated numbers need only be included to ensure that the totals align with the available supervising time (as in figures 6.1 and 6.2).

6.20 For timetabled spaces, some judgement will be required in considering if the frequency of use justifies the number of spaces. If the initial frequency seems too high (above 95%), the number of spaces may need to be increased. If the frequency of use is low (below 60%), some multi-use spaces might be used for a variety of activities. At this stage, therefore, a slightly different list of space types may be drawn up to include spaces which might accommodate, for example, food preparation and science experiments at different times in

Notes
1: Parent helpers, etc. are not included, on the assumption that only paid staff time is sufficiently reliable.
2: In this example:
In the infants' department: six teachers teach six classes for 34 periods per week on average = FTE of 5.83. Four NTAs teach for 57 periods per week = FTE of 1.63 and the head and deputy teach for an average of nine periods per week = FTE of 0.25, the total FTE being 7.71.
In the junior department: eight teachers teach eight classes for 38 periods per week = FTE of 7.6. Six NTAs teach for 96 periods per week = FTE of 2.4 and the head and deputy teach for 20 periods per week = FTE 0.5, the total FTE being 10.5.
In figures 5.1 and 5.2, the FTE number of teachers and NTAs in contact with the pupils can be seen to just exceed the calculated number of spaces required.

Appendix 1: A 5-11 Primary School Schedule

| SPACE TYPES | number of spaces calculated | | | maximum frequency of use | rounded number of spaces | actual frequency of Use | existing permanent spaces | new spaces required |
	infant	junior	total (a)	(b)	(c)	(a/c)		
class bases in basic teaching area	4.97	6.38	**11.35**	90%	**14**	81%	10	**4**
timetabled supplementary areas:								
mixed use practical	0.96	1.28	**2.24**	90%	**3**	75%	2	**1**
food	0.36	0.4	**0.76**	90%	**1**	76%	1	-
small group withdrawal	0.91	1.5	**2.41**	90%	**3**	80%	2	**1**
studio	0.23	0.3	**0.53**	90%	**1**	63%	-	**1**
hall	0.29	0.4	**0.69**	75%	**1**	69%	1	-
outdoors	(0.17)	(0.2)	(0.37)	-	-	-	-	-
TOTAL	7.72	10.26	**17.98**	-	23	-	16	**7**

Figure 6.3: step 3 for whole school based on total of calculated spaces for six infant and eight junior classes, from figures 6.1 and 6.2.

Figure 6.4: steps 4 and 5 for whole school based on area guidelines in section 3. *some withdrawn groups will use the library in this case, so only two small group rooms will be needed.

SPACE TYPES	number of spaces required	maximum group size (pupils)	average area of space (m²)	total teaching area (m²) (number x area)
basic teaching class bases	14	30	56	784
timetabled supplementary areas:				
mixed use practical	3	8	19	57
food	1	8	19	19
small group withdrawal*	2	4 or 8	10	20
studio	1	30	32	32
hall	1	30	140	140
other supplementary areas				
library (inc. 1 withdrawal)	1		26	26
TOTAL TEACHING AREA		-		1078
Total gross area with teaching area at 62% of gross				1739

the same space.

6.21 Figure 6.3 compares the rounded number of spaces for the enlarged number on roll to the number of existing spaces. There is a need for more timetabled spaces as well as more class bases.

Studios and Halls

6.22 With smaller schools, the demand for both a studio space for music and a hall for apparatus-based PE may be accommodated in one space. In the example, the demand in the 315 place school for studio and hall-based activities was sufficiently low for only one space to be needed. The increasing number on roll raises the demand such that a hall and a studio will now be desirable. One reason for this is that the available hours for the hall are less than other spaces as it is used for dining, so the preparation and clearing time for tables must be taken into account. The maximum frequency of use for the hall is therefore 75%, on the basis that it is unavailable for the periods before and after lunch, leaving 30 out of 40 periods per week available.

Step 4: Space Sizes

6.23 Finally, the area of each space must be decided, ideally on the basis of the largest group size that will normally use it and on the activities that will take place within it (figure 6.4). The maximum group sizes in that column have been judged from the likely upper limit of the infant and junior averages.

Step 5: Additional Areas

6.24 The information on timetabled spaces and their area can then be added to the other untimetabled supplementary teaching area, including the library, and a notional non-teaching area based on the guidance in section 3. Figure 6.4 illustrates the final schedule chosen in this example. The total teaching area is comfortably within the guidelines of gross area for a school of this size, by using 62% of the gross area as teaching area. In other cases, some adjustments may be required to individual spaces to address the overall use of all the area and its relationship to the acceptable gross area.

Final Adjustments

6.25 Depending on the acceptable gross area, based on the individual needs of the school, the following options are also available:

- increase the group sizes for some timetabled areas to decrease the demand (although this will increase the area required for some rooms);

- reduce the supplementary areas and allow for such activities to take place in the basic teaching area;

- reduce the basic teaching area to allow for all the preferred supplementary spaces to be accommodated;

- share the use of supplementary spaces such that the frequency of use of all spaces can be increased and less space can be used more effectively.

6.26 Figure 6.5 illustrates the second option listed above, within the same gross area as the chosen schedule.

6.27 Some final checks should be made on the completed schedule. It is important that:

- the total teaching area should be a reasonable percentage of the gross area available within the guidelines in Section 1, i.e. around 57% for a small school and at least 60% for a large one (as in this example);

- the classroom or equivalent accommodation provides the appropriate capacity when calculated using Annex D of Circular 6/91;

- the provision of timetabled spaces does not tend to reduce the use of the class bases to below about 80% of the working week (if the class bases are occupied for much less than this, they

could be partially timetabled to accommodate other activities);

- any specific timetabled spaces provided are in use for a reasonable proportion of the week;

- that sufficient and appropriate staff (i.e. teachers and NTAs) are likely to be available for supervision of the timetabled spaces, while retaining enough staff time in the classroom (see paragraph 6.17).

Other Options

6.28 If the teaching method of the school had been to use a more open plan approach, many of the discrete areas listed would be incorporated in the overall basic teaching area. Similarly, if the school preferred all but the hall and studio based activities to be in enclosed classrooms, the classes may be nearer $63m^2$ and resources would be duplicated in classrooms or moved around on trolleys (as in figure 6.5).

Figures 6.5: an alternative option for the final schedule of accommodation.
*this total area could comprise a series of 14 enclosed classrooms, as illustrated, or a mixture of 14 smaller bases each with access to shared teaching areas that might be used for practical activities.

ALTERNATIVE: ENCLOSED CLASSROOM DESIGN

SPACE TYPES	number of spaces required	maximum group size (pupils)	average area of space (m²)	total teaching area (m²) (number x area)
basic teaching area:				
classrooms (including practical)	14	30	63	882*
timetabled supplementary areas				
studio	1	30	40	30
hall	1	30	140	140
other supplementary areas				
library (inc. withdrawal)	1		16	26
TOTAL TEACHING AREA			-	1078
Total gross area with teaching area at 62% of gross				1739

Appendix 2: An 11-16 Secondary School Schedule

This appendix details the method, including curriculum analysis, used to create the schedule of accommodation for the 900 place 11-16 secondary school summarised in section 2. This example is a school with a relatively academic curriculum, within a gross area at the bottom of the range.

7.1 As with the primary school example in appendix 1, a gross area and teaching area range is estimated using the guidance in section 1. Then a curriculum analysis is used to identify the curriculum demand for timetabled spaces, as outlined in section 2. These areas, together with those of the other teaching spaces and non-teaching spaces are determined from the guidance in section 4 and are added to the schedule. Finally, adjustments are made to the overall schedule to ensure that it conforms to the original area constraints.

Overall Areas

7.2 Using the formulae in section 1, the gross area range for an 11-16 school of 900 is 6150m² to 6700m² in total[1]. In this example the lowest area will be the aim, with a teaching area at 60% of the gross area, amounting to 3690m². This teaching area will be predominantly timetabled teaching spaces but will also include

untimetabled teaching areas such as the library resource centre.

Curriculum Analysis

7.3 The five steps described in section 2 are followed in this example, the first four of which cover the curriculum analysis: using a statement of the curriculum to calculate the demand for timetabled spaces. In order to quantify the overall curriculum data, the following curriculum breakdown has been used.

Curriculum Breakdown

7.4 The curriculum breakdown in figure 7.2 shows one method of totalling the complexities of organisation of all five year groups in the school. Other methods could be used to equal effect. The aim is to identify the total pupil periods, teacher periods and average group size for each subject.

7.5 This example is somewhat simplified in that years 7, 8 and 9 have the same curriculum. The spread sheet in figure 7.2 would also serve if the data were different for each year group, for instance if they were different sizes (although the year group sizes are usually assumed to be the same for design purposes). In each of the

Figure 7.1: data used to calculate the pupil periods and teacher periods in years 7, 8 and 9.

DATA FOR YEARS 7,8,9				pupils in year =180 (aa)		
subject	periods each pupil taught	maximum group size	number of groups in year	pupil periods	teacher periods	average group size
	ab	ac	ad	ae	af	
			(aa/ac)	(aa x ab)	(ab x ad)	(ae/af)
English	5	27	7	900	35	25.7
maths	5	27	7	900	35	25.7
modern foreign languages	5	27	7	900	35	25.7
humanities	4	27	7	720	28	25.7
religious education	2	27	7	360	14	25.7
PSE	1	27	7	180	7	25.7
IT courses	1	20	9	180	9	20
science	5	27	7	900	35	25.7
design and technology	3	20	9	540	27	20
art	2	30	6	360	12	30
music	2	30	6	360	12	30
drama	1	30	6	180	6	30
PE (indoor)	2	30	6	360	12	30
games (outdoor)	2	30	6	360	12	30
TOTAL	**40**	-	-	**7200**	**279**	-

Notes
1: The total gross area, based on the formulae in section 1, will range from
(900 x 5.5) + 1200 = 6150m²
to
(900 x 6) + 1300 = 6700m²

60

11-16 Curriculum Breakdown

a	=	periods per week	40
b	=	total number on roll	900
c	=	FTE number of teachers	47.6
d	=	contact ratio	0.78
e	=	total pupil periods (a x b)	**36000**
f	=	total teacher periods (a x c x d)	**1485**
g	=	periods teachers teach (a x d)	31.2

Number of Pupils	year 7 180		year 8 180		year 9 180		year 10 180		year 11 180		TOTAL 900		FTE number of teachers
	pupil periods	teacher periods	pupil periods	teacher periods	pupil periods	teacher periods	pupil periods	teacher periods	pupil periods	teacher periods	total pupil periods	total teacher periods	
English	900	35	900	35	900	35	900	40	900	40	4500	185	5.9
mathematics	900	35	900	35	900	35	900	40	900	40	4500	185	5.9
modern Languages	900	35	900	35	900	35	720	32	720	32	4140	169	5.4
humanities	720	28	720	28	720	28	720	32	720	32	3600	148	4.7
religious education	360	14	360	14	360	14	360	16	360	16	1800	74	2.4
PSE	180	7	180	7	180	7	-	-	-	-	540	21	0.7
general studies	-	-	-	-	-	-	360	16	360	16	720	32	1.0
TOTAL GENERAL	3960	154	3960	154	3960	154	3960	176	3960	176	19800	814	26.1
IT/ business studies	180	9	180	9	180	9	180	9	180	9	900	45	1.4
science	900	35	900	35	900	35	1440	64	1440	64	5580	233	7.5
design and technology	540	27	540	27	540	27	540	27	540	27	2700	135	4.3
art	360	12	360	12	360	12	180	8	180	8	1440	52	1.7
music	360	12	360	12	360	12	180	8	180	8	1440	52	1.7
drama	180	6	180	6	180	6	180	8	180	8	900	34	1.1
P E (indoor)	360	12	360	12	360	12	270	12	270	12	1620	60	1.9
games (outdoor)	360	12	360	12	360	12	270	12	270	12	1620	60	1.9
TOTAL	7200	279	7200	279	7200	279	7200	324	7200	324	**36000**	**1485**	47.6
FTE number of teachers		8.9		8.9		8.9		10.4		10.4			
pupil:teacher ratio		20.1		20.1		20.1		17.3		17.3			18.9
special educational needs		15		12		12						0	39

first three years, the pupil periods and teacher periods are based on the data in figure 7.1.

7.6 The pupil periods stem from the number of periods each pupil needs to be taught each subject over the 40 period week (column ab, figure 7.1) multiplied by the number of pupils in the year group (aa in figure 7.1). The teacher periods are based on the maximum group size in which each subject will be taught. In this case, the school have agreed that 27 should be the maximum size of general teaching groups, so the year group of 180 will be divided into seven groups and the actual average group size will be:

$$180 \div 7 = 25.7$$

7.7 Sometimes (for instance for PE) the year group is divided into six, when the average group size is the standard form of entry size of 30. When the year is in nine groups, as with design and technology, the groups average 20.

7.8 As figure 7.1 shows, the teacher periods for each subject (column af) are the number of periods each pupil needs to be taught (column ab) multiplied by the number of concurrent groups (column ad).

7.9 The information for years 10 and 11 has been calculated on a similar basis, with an increased number of groups in the year for most subjects resulting in a lower average group size. One reason for this is to allow option groups to be created. It may be necessary to break KS4 information down further if the options available are complex.

7.10 The full time equivalent (FTE) of the number of teachers can be usefully checked by dividing the teacher periods for any year or subject by the number of periods that teachers teach (g in figure 7.2). The pupil:teacher ratio may also be calculated for each year by dividing the number of pupils in the year by the FTE number of teachers (in the shaded areas in figure 7.2).

Figure 7.2: curriculum breakdown for 900 place 11-16 school example.

Step 1: Distribution of Time

7.11 As illustrated in figure 7.3, the first step is to list the total pupil periods (column v) and teacher periods (column w) resulting from the curriculum breakdown discussed above (figure 7.2), and to calculate the average group size across the school by dividing the pupil periods by the teacher periods (column x). The overall percentage of pupil time spent on each subject can also be inserted (column u). This can be a useful check and, if the spread sheet is reconfigured, the percentage figures and group sizes alone can be used to estimate the demand for changing numbers on roll (see paragraph 2.22).

Figure 7.3: steps 1 and 2 of the curriculum analysis, plus the initial rounding of the number of spaces, in step 3.

7.12 As a check, the total number of pupil periods should equal the periods per week (a) multiplied by the total number on roll (b). Similarly, the total number of teacher periods should equal the periods per week (a) multiplied by the FTE number of teachers (c) multiplied by the contact ratio (d).

Different types of space in one subject

7.13 Some subjects, notably design and technology, require a number of different types of space. It is possible to work out the number of spaces required for the subject up to step 3, then decide what kinds of space they will be. It may be

a	=	periods per week	40
b	=	total number on roll	900
c	=	FTE number of teachers	47.6
d	=	contact ratio	0.78
e	=	total pupil periods (a x b)	36000
f	=	total teacher periods (a x c x d)	1485

11 - 16 pupils		STEP 1				STEP 2	STEP 3	
subject		percentage of curriculum	total teacher periods	total teacher periods	average group size	number of spaces calculated	number of spaces rounded	frequency of use
		u	v	w	x	y	z	
			from	from			nearest	y as
Method		v as	curriculum	curriculum	v/w	w/a	whole no.	percentage
Explanation		percentage	breakdown	breakdown			above y	of z
		of total	(figure 7.2)	(figure 7.2)				
TOTAL GENERAL		**55.0%** =	**19800**	**814**	-	**20.35**		
standard	@ 77.5%	=	15345	631	24.3	15.77	16)	96.9%
large	@ 22.5%	=	4455	183	24.3	4.58	5)	
IT/ business studies		**2.5%** =	**900**	**45**	**20.0**	**1.13**	2	56.3%
science		**15.5%** =	**5580**	**233**	**23.9**	**5.83**	6	97.1%
design and technology		**7.5%** =	**2700**	**135**	-	**3.38**		
food	@ 21.5%	=	581	29	20.0	0.73	1	72.6%
multi-mat/graphics	@ 35.5%	=	959	48	20.0	1.20	2	59.9%
PECT	@ 21.5%	=	581	29	20.0	0.73	1	72.6%
textiles	@ 21.5%	=	581	29	20.0	0.73	1	72.6%
art		**4.0%** =	**1440**	**52**	-	**1.30**		
2D art	@ 60.0%	=	864	31	27.7	0.78	1	78.0%
3D art/textiles	@ 40.0%	=	576	21	27.7	0.52	1	52.0%
music		**4.0%** =	**1440**	**52**	**27.7**	1.30	2	65.0%
drama		**2.5%** =	**900**	**34**	**26.5**	0.85	1	85.0%
P E (indoor)		**4.5%** =	**1620**	**60**	**27.0**	1.50	2	75.0%
games (outdoor)		**4.5%** =	**1620**	**60**	**27.0**	1.50	(external space)	
TOTAL		**100%**	**36000**	**1485**	**37.13**	**41**		

more accurate to identify the number and type of spaces needed by sub-dividing the teaching and pupil periods of the subject at step 1, as in figure 7.3. The proportions in column u for food, multi-materials/graphics, PECT and textiles are typical but may change depending on the school and its strengths. A similar sub-division has been used to identify the number of large general teaching rooms required and the need for a 3D art space including ceramics or wet textiles.

Step 2: Calculated Number of Spaces

7.14 Having established the information in step 1, the number of spaces demanded can be calculated in the next column (y) by dividing the number of teacher periods by the periods per week.

7.15 The proportions of time for specialist design and technology spaces, large and standard general teaching rooms, and 2D and 3D art have been multiplied by the total figures for the subject to give an exact calculated number of spaces required for each.

Step 3: Rounded Number of Spaces

7.16 At this point, the actual numbers and types of spaces required can be decided, based on a reasonable frequency of use, by rounding the calculated figures up. In figure 7.3, an initial rounding up has been done. By looking at the frequencies of use resulting from this first attempt, a more reasonable number can be estimated, as shown in figure 7.4.

7.17 In this simple example, very few changes have been made, but other examples may need more judgement in considering if the frequency of use justifies the number of spaces. If the initial frequency seems too high (above about 90% except in PE), the number of spaces may need to be increased. If the frequency of use is low (below 60% unless the space is used for free access as well), some rooms might be identified to cater

for the activities of more than one space type. At this stage, therefore, a slightly different list of space types may be drawn up.

7.18 In this case, the following considerations have been made:

- the number of standard general teaching spaces has been increased to 18, with five large general teaching rooms, giving a total of 23 at a more reasonable frequency of use of 88.5%;

- the two IT spaces have a low timetabled use at 56%, but will also be available for irregular bookings of classes in other subjects, or for small groups from other classes, so this was felt to be reasonable;

- the number of science laboratories has been increased to seven to reduce the usage to an acceptable 83%;

- 3D art is only to be used 52% of the time, but in practice this room will also be used for 2D art, so the frequency of use will even out between the two art rooms;

- some drama may be taught in one of the music rooms, and vice versa, to allow more flexibility in teaching and timetabling, which would mean the overall frequency of use for music and drama would be 72% (the average of all three spaces);

- PE would require two spaces at 75% usage each, or a sports hall. The assembly hall has been identified as the second PE space, used for up to 60% of the time for dance or movement. In practice, this may be less if outdoor PE facilities can be more heavily used.

Step 4: Space Sizes

7.19 Finally, the area of each space must be decided, ideally on the basis of the largest group size that will normally use it and on the activities that will take place within it. The maximum group sizes have been judged from the likely upper limit in the curriculum breakdown, bearing in mind that room sizes need to be designed to suit possible minor changes in group

Figure 7.4: steps 3 (continued), 4 and 5 of the curriculum analysis.

Types of Space	STEP 2 number of spaces calculated y	STEP 3 number of spaces adjusted z	STEP 3 frequency of use	STEP 4 maximum group size	STEP 4 average area per space m²	STEP 4 total teaching area m²	MOE WORKPLACES work-places per space	MOE WORKPLACES total work-places
method explanation		based on reasonable freq. of use	y as percentage of z	based on curriculum breakdown	from section 4 guidelines	z x average area	based on Circular 11/88	workplaces per space x no. of spaces
total general								
standard	15.77	**18**)	88.5%	30	50	**900**	30	540
large	4.58	**5**)		30	62	**310**	30	150
IT/business studies	1.13	**2**	56.3%	25	72	**144**	20	40
science	5.83	**7**	83.2%	30	85	**595**	30	210
design and technology								
food	0.73	**1**	72.6%	21	103	**103**	20	20
multi-mat/graphics	1.20	**2**	59.9%	21	103	**206**	20	40
PECT	0.73	**1**	72.6%	21	88	**88**	30	30
textiles	0.73	**1**	72.6%	21	84	**84**	30	30
art								
2D art	0.78	**1**	78.0%	30	91	**91**	30	30
3D art/textiles	0.52	**1**	52.0%	30	109	**109**	30	30
music	1.30	**2**	65.0%	30	70	**140**	20	40
drama	0.85	**1**	85.0%	30	91	**91**	20	20
PE (indoor)	1.50	**1**	90.0%*	30	260	**260**	30	30
games (outdoor)	1.50	(external space)						
TOTAL	**37.13**	**43**				**3121**		

STEP 5

NON-TIMETABLED SPACES

SEN withdrawal space		1		6	21	**21**	15	15
library resource centre		1		-	143	**143**	30	30
local/IT resource areas		2		15	28	**56**	15	30
careers area		1		5	13	**13**		
FLA/seminar space		1		3	7	**7**		
music group rooms		5		4	8	**40**		
darkroom		1		3	10	**10**		
heat treatment bay		1		4	15	**15**		
kiln		1		-	4	**4**		
assembly hall (*used for PE)		1	60.0%	-	260	**260**	30	30

TOTAL TEACHING AREA **3690**

NON-TEACHING AREA

staff accommodation	@ 5.5% of gross	338
pupils' storage/washrooms	@ 5.0% of gross	308
teaching storage	@ 5.0% of gross	308
catering facilities	@ 4.5% of gross	277
ancillary/circulation/partitions	@ 20.0% of gross	1230

TOTAL GROSS AREA **6150**   **1315**

lower line for gross area	6150 m2	
upper line for gross area	6700 m2	(teaching area = 60% of gross area)

sizes in the longer term. In this example, all areas are based on the middle of the ranges recommended in section 4.

Step 5: Additional Areas

7.20 The information on timetabled spaces and their area has then to be added to the other untimetabled supplementary teaching area, including the SEN support spaces (based on the amount of teaching requiring a special room for withdrawn groups). The overall area of the library resource centre, the local resource areas or IT clusters needed and the careers area, are from the middle of the range in figure 4.10.

7.21 A notional non-teaching area, based on 40% of the lowest gross area of the range, has been divided into five types of space with a guide figure for each based on the average of the guidance in section 4.

Final Adjustments

7.22 In this example, the acceptable gross area, based on the individual needs of the school, has been achieved by the measures described above. Within the same or a slightly larger gross area, a number of other options would be available, including:

- to include a sports hall, equivalent to two PE spaces at 75% frequency of use each[1], and use the assembly hall for drama. The remaining drama and music could then share two rooms (this option is shown in figure 2.1);

- use an art space as a graphics room for some design and technology and teach PECT in a multi-materials workshop, resulting in only four design and technology spaces;

- reduce the untimetabled teaching area to allow for more timetabled spaces to be accommodated, or vice-versa.

7.23 The final stage is completed by checking the following points:

Gross Area

7.24 As discussed, the final gross area falls at the lower end of the ranges outlined in section 1. The teaching area is 60% of the gross. In a smaller school or one with other constraints, a lower figure of 57% to 59% may be acceptable.

MOE Capacity Compatibility

7.25 The number of workplaces provided by this schedule can be calculated using the definitions in annexe A of Circular 11/88, as can be seen in the last two columns of figure 7.4. The capacity of the proposed schedule, using the 'MOE formula' method in annexe A of Circular 11/88, is 967. This is 7% to 8% over the number on roll for which the schedule has been planned, which allows for some variations in admission numbers each year and some pupils with special needs.

Average Frequency of Use

7.26 The FTE number of teachers in the school is 47.6 (as in figure 7.2) and the contact ratio is 0.78. The calculated number of timetabled spaces is 37.13 (47.6 x 0.78). The final number of spaces is 43, which results in a reasonable average frequency of use of timetabled spaces of around 86.3%.

Registration Bases

7.27 The basic organisation of this school is that the six forms of entry in each year are normally divided into seven for teaching classes. Several subjects are still taught in groups of 30 in years 7, 8 and 9 and in years 10 and 11 groups vary depending on options, so it may be reasonable to assume that 30 bases are required (six groups per year). In this brief, there are 23 general teaching rooms and the four art and music rooms are available, leaving three groups that will have to be registered elsewhere, such as in the library, IT/ business studies rooms or other specialist rooms not containing easily damaged or potentially hazardous equipment.

Note
1: In practice, a sports hall may be used by two teaching groups for some activities, such as gymnastics, and by one group for other areas of the curriculum, such as team games, so the frequency of use of the sports hall is likely to be higher than the 75% calculated for the equivalent two spaces.

Appendix 3: An 11-18 Secondary School Schedule

This appendix uses a curriculum analysis to create a schedule of accommodation similar to the 11-16 school in appendix 2, but with an additional sixth form. Also, in this example, the school has a more vocational curriculum, and aims to accommodate the demands of the brief within a gross area in the upper part of the range.

8.1 The 11-16 curriculum is similar to that in appendix 2, for ease of comparison. To emphasise a more vocational slant, the percentage of time pupils spend learning design and technology has been increased to 10%, at the expense of some RE teaching, and the group sizes have been reduced[1]. The demand for design and technology spaces therefore increases, while the need for general teaching rooms is reduced. Similarly, at KS4 art is taught more and IT is not studied as a separate course.

8.2 The post-16 curriculum contains GNVQ courses in business (both intermediate and advanced), advanced health and social care and science and one year intermediate courses in art and design and manufacturing, as well as a number of A-levels and general studies.

Overall Areas

8.3 This school has a stay-on rate of 60% on average, or 215 pupils in Years 12 and 13. Using the formulae in section 1, the gross area for an 11-18 school with a total number on roll of 1115 is shown to be in the range of 7870m² to 8635m² in total[2]. As the intention is to aim for an upper-range area, this might be around 8500m², with a teaching area, at 60% of the gross area, of 5070m² including timetabled and non-timetabled spaces.

Curriculum Analysis

8.4 The five steps outlined in section 2 are followed, as in the previous appendices. Like appendix 2, a curriculum breakdown has been used to determine the data required.

Curriculum Breakdown

8.5 The 11-16 and 16-18 curriculum breakdowns are done separately. In this example, the curriculum data in figure 8.1 is collated relating to a range of space-types (like appendix 1) as well as subjects. In figures 8.1 and 8.2, the overall teacher periods for all subjects or courses are totalled by subject in the bottom row, with the pupil periods, percentage of pupil time and the calculated number of spaces for each subject, to give an indication of the rooms required for each subject (for instance, for a suite of spaces). But they are also totalled under the different types of space required, in the right hand column. This demand for space-types is the data that are required for the schedule of accommodation, but the simple method of assuming all spaces are subject-specific can be difficult in 11-18 examples.

8.6 Courses such as GNVQs often need to be taught in a variety of specialist spaces, both practical and general. In figure 8.2, a number of examples of this can be seen.

- GNVQ business, like all GNVQ courses, involves some core units such as numeracy and communication, which may be timetabled in classrooms or in a lightly equipped GNVQ/business room. Other units may require a business office environment or heavy use of IT equipment.

- Some timetabled time may be away from the school site, in visits and work experience. Such activities are not usually regular enough to show on a weekly timetable, but in this example two periods per week in GNVQ health and social care are timetabled off site.

- GNVQ manufacturing is usually done with a range of practical facilities, including those in food, multi-materials and PECT rooms, while core units and related study may be done in a small general teaching space, an IT room or a GNVQ 'base'. These various activities may be possible in one or two multi-purpose rooms, but this method enables relatively standard space-types to be shared with other courses.

Notes
1: Teacher periods in bold in figure 8.1 indicate the changes to the example in appendix 2.
2: The total gross area, based on the formulae in section 1, will range from
(5.5 x 1115) +1200 +(2.5 x 215) = 7870m²
to
(6 x 1115) +1300 +(3 x 215) = 8635m²

Figures 8.1 and 8.2:
curriculum breakdown of 11-16 and 16-18 pupils by subject and activity-related space type.

11-16 CURRICULUM BREAKDOWN

Figure 8.1

	English teach group per / size	maths teach group per / size	modern languages teach group per / size	humanities teach group per / size	RE/PSE gen. studies teach group per / size	IT teach group per / size	science teach group per / size	design & technology teach group per / size	art teach group per / size	music teach group per / size	drama teach group per / size	PE teach group per / size	TOTAL teacher periods	average group size
year 7	35 25.7	35 25.7	35 25.7	28 25.7	14 25.7	8 22.5	35 25.7	36 20	12 30	12 30	6 30	24 30	280	25.71
year 8	35 25.7	35 25.7	35 25.7	28 25.7	14 25.7	8 22.5	35 25.7	36 20	12 30	12 30	6 30	24 30	280	25.71
year 9	35 25.7	35 25.7	35 25.7	28 25.7	14 25.7	8 22.5	35 25.7	36 20	12 30	12 30	6 30	24 30	280	25.71
year 10	40 22.5	40 22.5	32 22.5	32 22.5	24 22.5		64 22.5	40 18	16 22.5	8 22.5	8 22.5	24 22.5	328	21.95
year 11	40 22.5	40 22.5	32 22.5	32 22.5	24 22.5		64 22.5	40 18	16 22.5	8 22.5	8 22.5	24 22.5	328	21.95
TOTAL/AVERAGE	185 24.3	185 24.3	169 24.5	148 24.3	90 24	24 22.5	233 23.9	188 19.1	68 26.5	52 27.7	34 26.5	120 27	1496	24.06
TYPES OF SPACE:														
General Teaching														
small group														
standard group	175 24.3	143 24.3	169 24.5	52 24.3	55 24								594	24.34
large space	10 24.3	42 24.3		96 24.3	25 24								173	24.27
Practical Spaces														
GNVQ/business														
IT room					10 24	24 22.5							34	22.94
science lab: small group														
science lab: standard group							233 23.9						233	23.94
food @ 28.0%								50 19.1					50	19.15
multi-materials @ 28.0%								58 19.1					58	19.15
PECT @ 22.0%								40 19.1					40	19.15
textiles @ 17.0%								30 19.1					30	19.15
graphics/CAD @ 5.0%								10 19.1					10	19.15
2D art @ 65.0%									44 26.5				44	26.47
3D art/textiles @ 35.0%									24 26.5				24	26.47
music room										52 27.7			52	27.69
drama studio											34 26.5		34	26.47
PE spaces														
gym/sports hall @ 50.0%												60 27	60	27
off-site/outside @ 50.0%												60 27	60	27
TOTAL	185 24	185 24	169 24	148 24	90 24	24 23	233 24	188 19	68 26	52 28	34 26	120 27	1496	
total pupil periods	4499	4499	4139	3599	2159	540	5579	3600	1800	1440	900	3240	35992	
percentage of curriculum	12%	12%	11%	10%	6%	2%	15%	10%	5%	4%	3%	9%		
calculated number of spaces	4.63	4.63	4.23	3.7	2.25	0.6	5.83	4.7	1.7	1.3	0.85	3		

POST-16 CURRICULUM BREAKDOWN

Figure 8.2

	NOR per year	English teach group per / size	maths teach group per / size	modern languages teach group per / size	economics/ humanities teach group per / size	science teach group per / size (incl. GNVQ)	IT/ business teach group per / size (incl. GNVQ)	health & social care teach group per / size (GNVQ)	manufac-turing teach group per / size (GNVQ Y12)	art & design teach group per / size (GNVQ Y12)	music teach group per / size	PE/sports studies teach group per / size	general studies teach group per / size	TOTAL teacher periods	average group size
one year course															
year 12	45	9 15	9 15				22 21		22 12	22 12		6 15	6 15	96	15
two year courses															
year 12	85	16 15	20 15	16 12	32 13	44 13	20 15	20 10		8 10		14 15.6	10 17	200	13.44
year 13	85	16 15	20 15	16 12	32 13	44 13	20 15	20 10		8 10	8 6	14 15.6	10 17	208	13.15
TOTAL/AVERAGE	215	41 15	49 15	32 12	64 13	88 13	62 17.1	40 10	22 12	38 11.2	8 6	34 15.5	26 16.5	504	13.62
TYPES OF SPACE:															
general teaching															
small group		41 15	49 15	32 12	60 13							4 6	6 15	192	13.69
standard group													10 17	10	17
large space															
practical spaces															
GNVQ/business						12 13	37 17.1	24 10	6 12	6 11.2				85	13.75
IT room					4 13		25 17.1	10 10					4 17	43	15.08
science lab: small group						76 13						4 6		80	12.65
science lab: standard group															
food								4 10	6 12					10	11.2
multi-materials									4 12					4	12
PECT									4 12					4	12
textiles															
graphics/CAD									2 12	4 11.2				6	11.44
2D art @ 70.0%										18 11.2			2 17	20	11.74
3D art/textiles @ 30.0%										8 11.2				8	11.16
music room											8 6		2 17	10	8.2
drama studio													2 17	2	17
PE spaces															
gym/sports hall @ 50.0%												13 18.4		13	18.4
off-site/outside @ 50.0%								2 10				13 18.4		15	17.28
Self-study														(35)	(50)
TOTAL		41 15	49 15	32 12	64 13	88 13	62 17	40 10	22 12	36 11	8 6	34 15	26 17	502	
total pupil periods		615	735	384	832	1144	1062	400	264	402	48	526	430	6842	
percentage of curriculum		9%	11%	6%	12%	17%	16%	6%	4%	6%	1%	8%	6%		
calculated number of spaces		1.03	1.23	0.8	1.6	2.2	1.55	1	0.55	0.9	0.2	0.85	0.65		

Appendix 3: An 11-18 Secondary School Schedule

Figure 8.3: steps 1 and 2 of the curriculum analysis, plus the initial rounding of the number of spaces, in step 3, for the 11-16 and post-16 pupils separately.

BASIC DATA

a	=	Periods per week		40
c	=	FTE number of teachers	FTE	66.6
d	=	Contact Ratio		0.75
f	=	Total Teacher Periods (a x c x d)	1998	= total w for whole school

11-16 CURRICULUM ANALYSIS

b_1	=	Total 11-16 Number on Roll	900	
e_1	=	Total 11-16 Pupil Periods (a x b1)	36000	= total v for 11-16

	STEP 1					STEP 2	STEP 2	
TYPES OF SPACE:	total teacher periods	approx. group size	approx. pupil periods	recalc'ed pupil periods	recalc'ed group size	number of spaces calculated	number of spaces rounded	frequency of use
	w	p	q	v	x	y_1	z_1	
Method Explanation	from curriculum breakdown (figure 8)	from curriculum breakdown	w x p	q x (total q)/e1	v / w	w / a	nearest whole no. above y_1	y as percentage of z_1
General Teaching								
small group	0							
standard group	594	24.3	14455	14458	24.3	14.85	15	99.0%
large space	173	24.3	4199	4200	24.3	4.33	5	86.5%
Practical Spaces								
GNVQ/business								
IT room	34	22.9	780	780	22.9	0.85	1	85.0%
science lab: small group	0							
science lab: standard group	233	23.9	5579	5580	23.9	5.83	6	97.1%
food	50	19.1	957	958	19.2	1.25	2	62.5%
multi-materials	58	19.1	1111	1111	19.2	1.45	2	72.5%
PECT	40	19.1	766	766	19.2	1.00	2	50.0%
textiles	30	19.1	574	575	19.2	0.75	1	75.0%
graphics/CAD	10	19.1	191	192	19.2	0.25	1	25.0%
2D art	44	26.5	1165	1165	26.5	1.10	2	55.0%
3D art/textiles	24	26.5	635	635	26.5	0.60	1	60.0%
music room	52	27.7	1440	1440	27.7	1.30	2	65.0%
drama studio	34	26.5	900	900	26.5	0.85	1	85.0%
PE spaces								
gym/sports hall	60	27.0	1620	1620	27.0	1.50	2	75.0%
off-site/outside	60	27.0	1620	1620	27.0	1.50	2	75.0%
TOTAL	**1496**		**35992**	**36000**		**37.40**	**45**	**83.1%**

POST-16 CURRICULUM ANALYSIS

b_2	=	Total post-16 Number on Roll	215	
e_2	=	Total post-16 Pupil Periods (a x b)	8600	= total v for post-16

	STEP 1					STEP 2	STEP 2	
TYPES OF SPACE:	Total Teacher Periods	approx. group size	approx. pupil periods	recalc'ed pupil periods	recalc'ed group size	number of spaces calculated	number of spaces rounded	Frequency of Use
	w	p	q	v	x	y_2	z_2	
Method Explanation	from curriculum breakdown (figure 8)	from curriculum breakdown	w x p	q x (total q)/e2	v / w	w / a	nearest whole no. above y_2	y as percentage of z_2
General Teaching								
small group	192	13.7	2628	2630	13.7	4.80	5	96.0%
standard group	10	17.0	170	170	17.0	0.25	1	25.0%
large space	0							
Practical Spaces								
GNVQ/business	85	13.7	1169	1170	13.8	2.13	3	70.8%
IT room	43	15.1	648	649	15.1	1.08	2	53.8%
science lab: small group	80	12.7	1012	1013	12.7	2.00	3	66.7%
science lab: standard group	0							
food	10	11.2	112	112	11.2	0.25	1	25.0%
multi-materials	4	12.0	48	48	12.0	0.10	1	10.0%
PECT	4	12.0	48	48	12.0	0.10	1	10.0%
textiles	0							
graphics/CAD	6	11.4	69	69	11.4	0.15	1	15.0%
2D art	20	11.7	235	235	11.8	0.50	1	50.0%
3D art/textiles	8	11.2	89	89	11.2	0.20	1	20.0%
music room	10	8.2	82	82	8.2	0.25	1	25.0%
drama studio	2	17.0	34	34	17.0	0.05	1	5.0%
PE spaces								
gym/sports hall	13	18.4	239	239	18.4	0.33	1	32.5%
off-site/outside	15	17.3	259	259	17.3	0.38	1	37.5%
Self-study	(35)	50.0	1750	1752	50.0	(0.88)	1	87.5%
TOTAL	**502**		**8592**	**8600**		**12.55**	**25**	**50.2%**

Total w for whole school = 1998   (pupil:teacher ratio @ 66.6 teachers = 16.7)

- Pupils studying A-level sports studies may need access to a science laboratory for lessons in human biology. If these can be timetabled, the demand can be shown on the curriculum breakdown.

- General studies often involves a number of activities, requiring time in a variety of types of space.

8.7 The demand for facilities such as IT for GNVQs may be difficult to achieve if each course is treated separately, but identifying the overall demand can lead to efficiently used, shared specialist spaces.

8.8 There is also a further breakdown of some spaces for the range of group sizes, as sixth form courses (and some KS4 subjects) often have sufficiently small groups to justify rooms for a maximum of, say, 16. Group sizes are usually relatively small for all years in practical subjects other than science, so only general teaching spaces and science laboratories are broken down into rooms for small and standard groups. The GNVQ/business rooms are assumed to house post-16 groups of 20.

Step 1: Distribution of Time

8.9 For the purposes of calculating the need for types of spaces, the total number of teacher periods and overall average group sizes for each type of space can be used, from the tables in the right hand column of the curriculum breakdowns (figures 8.1 and 8.2).

8.10 At this stage, a quick check of the data may be worthwhile. The total teacher periods should equal the FTE number of teachers x the contact ratio x the periods in a week. Similarly, the pupil-periods for each space-type, derived from the number of teacher periods x the average group size, should total the number on roll for each age range x the periods in the week.

8.11 In this example, this is an opportunity to reconfigure the average group sizes, as the original estimate may not be exact (as numbers cannot easily be predicted in the sixth forms). The shaded areas in figure 8.3 show the initial calculation. The reconfigured data is then altered slightly in proportion to the calculated total pupil periods and the actual number (the number on roll x the periods per week).

8.12 The self-study time of sixth form pupils needs to be accounted for, in order to ensure the pupil periods total is correct. In this example, private study has been given a notional number of teacher periods, not included in the total, which can be used to show the demand for a self-study area. An overall proportion of time that all sixth formers are expected to be untimetabled could also be used.

Step 2: Calculated Number of Spaces

8.13 In this example, the number of spaces for the 11-16 pupils and the sixth form have been calculated separately, then added together (figure 8.3). This can be useful if the sixth form has a longer week and more periods than the rest of the school.

8.14 It is also useful, if step 3 is done for each as in figure 8.3, to check the demand for sixth form use when calculating the final number and size of spaces in steps 3 and 4. For instance, if a total of four IT rooms were chosen, two would be predominantly used by KS3 and 4 pupils, in an average group size of 22.5, and two by sixth form groups of around 15. This could affect the area of the rooms, decided at step 4, and the furniture layouts.

Step 3: Rounded Number of Spaces

8.15 Having derived the total calculated number of spaces for each space-type and added the 11-16 and post-16 figures together, the actual number of each type of space can be decided (figure 8.4). A reasonable frequency of use will allow many subjects to have suites of subject-specific rooms, while other specialist spaces will be shared.

Appendix 3: An 11-18 Secondary School Schedule

Figure 8.4: step 3, 4 and 5 of the curriculum analysis for the whole school.

types of space	STEP 2 11-16 y_1	STEP 2 post-16 y_2	STEP 2 TOTAL y	STEP 3 number of spaces adjusted z	STEP 3 frequency of use	STEP 4 maximum group size	STEP 4 average area per space m^2	STEP 4 total teaching area m^2	MOE workplaces per space	MOE total workplaces
method explanation				based on reasonable freq of use	y as percentage of z	(bold indicates 6th form)	from section 4 guidelines	z x average area	based on circular 11/88	workplaces per space x no. spaces
general teaching										
small group	-	4.80	4.80	5)		16	41	**205**	20	100
standard group	14.85	0.25	15.10	17)	89.7%	30	50	**850**	30	510
large space	4.33	-	4.33	5)		30	62	**310**	30	150
practical spaces										
GNVQ/business studies	-	2.13	2.13	3	70.8%	20	75	**225**	20	60
IT room	0.85	1.08	1.93	3	64.2%	25	72	**216**	20	60
science lab: small group	-	2.00	2.00	3	66.7%	16	75	**225**	20	60
science lab: standard group	5.83	-	5.83	7	83.2%	30	85	**595**	30	210
food	1.25	0.25	1.50	2	75.0%	20	100	**200**	20	40
multi-materials	1.45	0.10	1.55	2	77.5%	20	100	**200**	20	40
PECT	1.00	0.10	1.10	1.5	73.3%	20	87	**131**	20	30
textiles	0.75		0.75	1	75.0%	20	82	**82**	20	20
graphics/CAD	0.25	0.15	0.40	0.5	80.0%	20	78	**39**	20	10
2D art	1.10	0.50	1.60	2	80.0%	30	91	**182**	30	60
3D art/textiles	0.60	0.20	0.80	1	80.0%	30	109	**109**	30	30
music room	1.30	0.25	1.55	2	77.5%	30	74	**148**	20	40
drama studio	0.85	0.05	0.90	1	90.0%	30	91	**91**	20	20
PE spaces										
gym/sports hall	1.50	0.33	1.83	2	76.0%*	30	260	**520**	30	60
off-site/outside	1.50	0.38	1.88	-	-	-	-			
self-study		(0.88)	(0.88)	(1)	87.5%	50	74	**74**	30	30
TOTAL	37.40	12.55	49.95	58				4402		

STEP 5

	number of spaces	frequency	maximum group size	average area per space	total teaching area	workplaces per space	total workplaces
NON-TIMETABLED SPACES							
SEN support space	1		7	21	**21**	15	15
sixth form common area	1		50	80	**80**	30	30
library resource centre	1		-	165	**165**	30	30
local/IT resource areas	2		15	36	**72**	5	10
careers area	1		6	14	**14**		
FLA seminar space	1		4	8	**8**		
music group rooms	6		4	8	**48**		
darkroom	1		3	10	**10**		
heatbay	1		4	16	**16**		
kiln	1		-	4	**4**		
assembly hall (*used for PE)	1	40.0%	-	260	**260**	30	30
TOTAL TEACHING AREA					5100		

NON-TEACHING AREA

staff accommodation	@ 5.5%	average proportion of gross		468
pupils storage/washrooms	@ 5.0%	"		425
teaching storage	@ 5.0%	"		425
catering facilities	@ 4.5%	"		383
ancillary/circulation/partitions	@ 20.0%	"		1700

TOTAL GROSS AREA		8500	1670
lower limit of gross area	7870	m^2	
upper limit of gross area	8635	m^2	

8.16 In this example, the number of each type of space is usually based on a simple rounding up to the nearest whole number above the calculated figure. Other examples may need more judgement in considering if the frequency of use justifies the number of spaces, as discussed in appendix 2 (paragraph 7.17). In this case, the following considerations have been made:

- The number of standard general teaching spaces has been adjusted to 17, with five large general teaching rooms and five small sixth form seminar rooms for groups of up to 16, giving a total of 27 at a reasonable frequency of use of 89.7%.

- The three GNVQ/business rooms have a low frequency of timetabled use at around 71%, but will also be available for GNVQ students in untimetabled periods (reducing the area needs for the self-study space).

- Similarly, the three IT rooms have a low timetabled use at about 65%, but will also be available for irregular bookings of classes in other subjects. Two untimetabled IT clusters are included for small groups from KS3 and 4 classes or sixth-formers in self-study periods.

- There are seven general science laboratories and three advanced laboratories for sixth form lessons, with an overall frequency of use of 78%.

- Graphics has been identified as a separate activity under KS3 and 4 design and technology (figure 8.1) and a specialist graphics space is timetabled for GNVQ art and design and manufacturing. Together they show a demand for one space used 40% of the time, which is unacceptably low. This room will therefore double as a PECT room, as two spaces used only for this would also have a low usage. The average frequency of use of the two reconfigured rooms will be about 75%.

- Some drama may be taught in a music room, and vice versa, to allow more flexibility in teaching and timetabling, which would mean the overall frequency of use in the music and drama rooms would be around 82%.

- The assembly hall is of a suitable size to be used for PE for up to 40% of the time. This reduces the usage of the sports hall to be the equivalent of two spaces at 76%, allowing it to be used by one group in some periods and by two in others (see paragraph 4.61).

- The notional teacher periods and group size for sixth form self-study (or private study periods), in figure 8.2, are not included in the totals, but indicate a demand for a study space. This may be linked with the library resource centre. The pupil periods derived for self-study indicate that sixth-formers will spend an average of 20% of timetabled time in self-study (1752 out of 8600).

Step 4: Space Sizes

8.17 Finally, the area of each space must be decided, ideally on the basis of the largest group size that will normally use it and on the activities that will take place within it. The maximum group sizes have been judged from the likely upper limit in the curriculum breakdown, bearing in mind that room sizes need to be designed to suit possible minor changes in group sizes in the longer term and that group sizes in sixth forms are less predictable and can be quite wide-ranging. In this case, sixth form groups are envisaged to be a maximum of 16, so the areas of seminar rooms and advanced science laboratories normally used only by the sixth form are based on this number (with a 50% increase for post-16 workplaces as post-16 group sizes equal $G \div 1.5$ in the formulae).

Step 5: Additional Areas

8.18 The information on timetabled spaces and their area has then to be added to the other untimetabled supplementary teaching area, including the SEN areas (based on the amount of teaching requiring a special room for withdrawn groups). The overall area of the library resource centre, the local resource areas or IT clusters needed and the careers area, are within the range in figure 4.10.

8.19 A notional non-teaching area, based on 40% of the lowest gross area limit, has been divided into five types of space with a guide figure for each based on the average of the guidance in section 4.

Final Adjustments

8.20 The final stage is completed by checking the following points:

Gross Area

8.21 As discussed, the final gross area falls towards the upper end of the ranges outlined in section 1. The teaching area is 60% of the gross.

MOE Capacity Compatibility

8.22 The proposed schedule provides 1670 workplaces, including the dining area, as can be seen in the last two columns of figure 8.4. These figures can be used to calculate the capacity, using the 'MOE formula' method in annexe A of Circular 11/88. In this case the capacity is 1193 if no pupils with special needs are allowed for, assuming that PECT is taken as a 'light craft' space and that the dining area is not used for teaching. This is sufficiently close to be acceptable as the MOE formula does not have a direct relationship to the operational assumptions: some allowance needs to be made for variations in admissions and special needs.

Average Frequency of Use

8.23 The FTE number of teachers in the school is 66.6 (as in figure 8.3) and the number of timetabled spaces is 58. The number of spaces is 87% of the number of teachers.

Registration Bases

8.24 As in appendix 2 (paragraph 7.27), the basic organisation of this school is six forms of entry in each year, so 30 registration bases are required (six groups per year). The sixth form will register, if necessary, in the common area or study space provided. There are 22 classrooms large enough for groups of 30 and the five art and music rooms are available, leaving three groups that will have to be registered elsewhere, such as in seminar rooms, the textiles room or other specialist rooms not containing easily damaged or potentially hazardous equipment.

Appendix 4: Summary of Area Formulae

This appendix summarises the formulae for the gross area of school buildings and site area for school grounds, both as a total and per pupil, for those types of schools already covered in the text and also for other types of schools. It uses the methods described for gross areas in paragraphs 1.21, and for site areas in paragraph 5.31. Although the formulae will still be a useful guide, primary and middle schools with significantly less than 90 or more than 630 on roll, and secondary schools with less than 600 pupils, will need to make individual judgements based on possible economies of scale. Each pair of formulae represents the range.

Note: The upper age in the age ranges shown is one year above the average age of the upper year group, so, for instance, junior schools take pupils aged 7,8,9 and 10. At 11, they join a secondary school. Schools with age ranges from 5 are assumed to include reception classes and may also include nursery pupils.

type of school with age range and number of year groups in each key stage (KS)	total gross area of school buildings	gross area per pupil	total site area of school grounds	site area per pupil
Infants age 5 - 7 (KS 1:3)	$200+3.8N$ $170+3.4N$	$3.8+200/N$ $3.4+170/N$	$3500+21N$ $2800+17.5N$	$21+3500/N$ $17.5+2800/N$
First age 5 - 8 (KS1:3, KS2:1)	$200+3.8N$ $170+3.4N$	$3.8+200/N$ $3.4+170/N$	$3588+27.1N$ $2887+23.6N$	$27.1+3588/N$ $23.6+2887/N$
First age 5 - 9 (KS1:3, KS2:2)	$200+3.8N$ $170+3.4N$	$3.8+200/N$ $3.4+170/N$	$3640+30.8N$ $2940+27.3N$	$30.8+3640/N$ $27.3+2940/N$
Primary age 5 - 11 (KS1:3, KS2:4)	$200+3.8N$ $170+3.4N$	$3.8+200/N$ $3.4+170/N$	$3700+35N$ $3000+31.5N$	$35+3700/N$ $31.5+3000/N$
First and Middle (combined) age 5 - 12 (KS1:3, KS2:4, KS3:1)	$263+4.04N$ $224+3.63N$	$4.04+263/N$ $3.63+224/N$	$4238+39.4N$ $3375+35.8N$	$39.4+4238/N$ $35.8+3375/N$
Junior age 7 - 11 (KS2:4)	$200+3.8N$ $170+3.4N$	$3.8+200/N$ $3.4+170/N$	$3850+45.5N$ $3150+42N$	$45.5+3850/N$ $42+3150/N$
Middle deemed primary age 8 - 12 (KS2:3, KS3:1)	$325+4.28N$ $277+3.85N$	$4.28+325/N$ $3.85+277/N$	$4888+51.6N$ $3862+48N$	$51.6+4888/N$ $48+3862/N$
Middle deemed secondary age 9 - 13 (KS2:2, KS3:2)	$450+4.75N$ $385+4.3N$	$4.75+450/N$ $4.3+385/N$	$5925+57.8N$ $4575+54N$	$57.8+5925/N$ $54+4575/N$
Secondary (lower) age 11 - 14 (KS3:3)	$700+5.7N$ $600+5.2N$	$5.7+700/N$ $5.2+600/N$	$16000+60N$ $14000+56N$	$60+16000/N$ $56+14000/N$
Secondary age 11 - 16 (KS3:3, KS4:2)	$1300+6N$ $1200+5.5N$	$6+1300/N$ $5.5+1200/N$	" "	" "
Secondary age 11 - 18 (KS3:3, KS4:2, 16+:2)	$1300+6N+3n$ $1200+5.5N+2.5n$	$6+1300/N+3n/N$ $5.5+1200/N+2.5n/N$	" "	" "
Secondary age 12 - 18 (KS3:2, KS4:2, 16+:2)	$1450+6.08N+3n$ $1350+5.58N+2.5n$	$6.08+1450/N+3n/N$ $5.58+1350/N+2.5n/N$	" "	" "
Secondary age 13 - 18 (KS3:1, KS4:2, 16+:2)	$1700+6.2N+3n$ $1600+5.7N+2.5n$	$6.2+1700/N+3n/N$ $5.7+1600/N+2.5n/N$	" "	" "
Secondary (upper) age 14 - 18 (KS4:2, 16+:2)	$2200+6.45N+3n$ $2100+5.95N+2.5n$	$6.45+2200/N+3n/N$ $5.95+2100/N+2.5n/N$	" "	" "

Bibliography

GENERAL

The Education (School Premises) Regulations 1996 HMSO 1996. ISBN 0 11 054152 9

Department for Education and Employment Circular 10/96: The 1996 School Premises Regulations 1996

Department of Education and Science Building Bulletin 7: Fire and the design of educational buildings (sixth edition) HMSO 1992. ISBN 0 11 270585 5

TITLES REFERENCED IN TEXT

Department of Education and Science Design Note 34: Area Guidelines for Secondary Schools 1983 (now superseded by this document)

Department for Education Building Bulletin 77: Designing for pupils with special educational needs: Special schools HMSO 1992. ISBN 0 11 270796 3

Department of Education and Science Building Bulletin 61: Designing for pupils with special educational needs: ordinary schools HMSO 1984. ISBN 0 11 270313 5*

Department for Education Making IT Fit (video and booklet) 1995

Department of Education and Science A&B Paper 15: Lockers and Secure Storage 1990

Department of Education and Science Building Bulletin 58: Storage of pupils' personal belongings HMSO 1980. ISBN 0 11 270494 8*

Department of Education and Science Building Bulletin 56: Nursery Education in converted space HMSO 1978. ISBN 0 11 270460 3*

Department of Education and Science Design Note 1: Building for Nursery Education 1968

Department of Education and Science Broadsheet 1: Nursery Education: low cost adaptation of spare space in primary schools 1980

Department for Education and Employment Building Bulletin 80: Science Accommodation in Secondary Schools: a Design Guide HMSO 1995. ISBN 0 11 270873 0

Department for Education and Employment Building Bulletin 81: Design and Technology Accommodation in Secondary Schools: a Design Guide HMSO 1995. ISBN 0 11 270917 6

The Sports Council Handbook of Sports and Recreational Building Design, volume 2: Indoor Sports (second edition) Butterworth Architecture 1995. ISBN 0 7506 1294 0

Department for Education and Employment Our School - Your School: community use of schools, after school activities 1995. ISBN 0 85 522473 8

HEALTH AND SAFETY

British Standards Institution BS 4163: Code of practice for health and safety in workshops of schools and similar establishments BSI 1984.

The workplace (health, safety and welfare) regulations 1992, SI 1992 No 3004 HMSO 1992. With:

> *Health and Safety Commission* Workplace health, safety and welfare: Workplace (Health, Safety and Welfare) regulations 1992: approved code of practice and guidance HMSO 1992

Department for Education and Employment A Guide to Safe Practice in Art & Design HMSO 1995. ISBN 0 11 270896 X

The Food Safety Act 1990 HMSO 1990.

CLEAPPS School Science Service Risk Assessment for Technology in Secondary Schools. CLEAPPS School Science Service 1990

The control of substances hazardous to health regulations 1988, SI 1988 No 1657. Updated 1994, SI 1994 No 3246 HMSO 1988 and 1994

Health and Safety Executive Occupational Exposure Limits 1984: Guidance Note EH 40 HMSO 1984

Health and Safety (display screen equipment) regulations 1992, SI 1992 No 2792 HMSO 1992

SERVICES

Department of Education and Science Design Note 17: Guidelines for environmental design and fuel conservation in educational buildings HMSO 1981 (to be revised)

The Electricity at work regulations 1989, SI 1989 No 635 HMSO 1989

British Gas and Department for Education and Science Guidance Notes on Gas Safety in Educational Establishments British Gas 1989

British Standards Institution BS 5925: Code of practice for ventilation principles and designing for natural ventilation BSI 1991

Health And Safety Executive Guidance Note GS23: Electrical Safety in Schools HSE 1985

British Standards Institution BS 8313: Code of practice for accommodation of building services in ducts BSI 1989

Institution of Electrical Engineers Regulations for electrical installations: IEE wiring regulations IEE 1991

Department of Education and Science Building Bulletin 70: Maintenance of mechanical services HMSO 1990. ISBN 0 11 270717 3

SCHOOL GROUNDS

The Sports Council **Handbook of Sports and Recreational Building Design, Volume 1 Outdoor Sports (second edition)** Butterworth Architecture 1993. ISBN 0 75 061293 2

Department of Education and Science **Building Bulletin 71: The Outdoor Classroom** HMSO 1990. ISBN 0 11 270730 0

Department of Education and Science **Broad Sheet 17: Playing fields: marginal areas** 1983

Department of the Environment and The Welsh Office **Planning Permission: A Guide for Business** 1994

Department of National Heritage **Sport: Raising the Game** 1995

Learning Through Landscapes **Special Places; Special People: The Hidden Curriculum of School Grounds** 1994. ISBN 0 94 761348 X

Learning Through Landscapes **Using School Grounds as an Educational Resource** 1990. ISBN 1 87 286504 6

Learning Through Landscapes **Play, Playtime and Playgrounds** 1992. ISBN 1 87 286510 0

Learning Through Landscapes **Ecology in the National Curriculum** 1990. ISBN 1 87 286502 X

Learning Through Landscapes **Trees in the School Grounds** 1992. ISBN 1 85 741095 5

Learning Through Landscapes **Pond Design Guide for Schools** 1988. ISBN 1 87 065122 7

Department of Education and Science **Design Note 11: Chaucer Infant and Nursery School, Ilkeston** 1973

Department of Education and Science **Design Note 47: St John's School, Sefton: the design of a new primary school** 1989

Department of Education and Science **Building Bulletin 48: Maiden Erlegh Secondary School** HMSO 1973. ISBN 0 11 270084 5*

Department of Education and Science **Building Bulletin 49: Abraham Moss Centre, Manchester** HMSO 1973. ISBN 0 11 270345 3*

Department of Education and Science **Building Bulletin 53: Guillemont Junior School, Farnborough, Hampshire** HMSO 1976. ISBN 0 11 270351 8*

Department of Education and Science **Building Bulletin 59: The Victoria Centre, Crewe: school and community provision in urban renewal** HMSO 1981. ISBN 0 11 270525 1*

National Children's Play and Recreation Unit **Playground Safety Guidelines** 1992. ISBN 0 85 522405 3

National Playing Fields Association **Facilities for Athletics** 1980. ISBN 0 90 085895 8

National Playing Fields Association **Gradients for sports facilities** 1983 (republished as TAN 23 1996)

National Playing Fields Association **Sports Turf** 1993. ISBN 0 49 149503

National Playing Fields Association **Schools Play Grounds** 1990. ISBN 0 94 698529 3

Football Association and The Sports Council **Artificial Grass Surfaces for Association Football** 1995. ISBN 0 90 097196 4X

The Sports Council **Artificial Turf for Hockey** 1990. ISBN 0 90 657798 5

A Guide to Habitat Creation Packard Publishing Limited 1991. ISBN 1 85 341031 4

FORTHCOMING PUBLICATIONS

Further guidance in the form of building bulletins is planned to cover the following areas:

school grounds;

nursery provision;

boarding accommodation;

accommodating art in secondary schools;

accommodating music in secondary schools;

GNVQ and sixth form accommodation;

large spaces.

* out of print but available from HMSO Books (Photocopies Section).

Glossary

General

CAD-CAM: Computer aided design for computer assisted machinery.

CD-ROM: Compact Disc Read Only Memory. Computerised reference data.

CNC: Computer Numerically Controlled.

FREQUENCY OF USE: The average amount of timetabled time that a space is used, expressed as a percentage of the total time available in the timetabled week.

FTE: Full Time Equivalent.

GROSS AREA: The total floor area of a building or buildings, measured to the inside face of external walls, including the area of internal walls.

GROUP SIZE (G): The number of pupils in a teaching group using a teaching space.

IT: Information Technology (this includes computer hardware and software).

KEY STAGE (KS): The periods in each pupils education to which the elements of the National Curriculum will apply. There are four key stages, normally related to the age of the majority of the pupils in a teaching group. They are: beginning of compulsory education to age 7 (KS 1); 7-11 (KS2); 11-14 (KS3) and 14 to end of compulsory education (KS4).

LEA: Local Education Authority.

m²: Square metres.

MOE: More Open Enrolment.

NC: National Curriculum.

NTA: Non-Teaching Assistant.

NUMBER ON ROLL (N): The total number of pupils enrolled in a school, including those in the sixth form. This may be assumed to be the same as the 'normal number' as defined in the School Premises Regulations[1].

PE: Physical Education.

PSE: Personal and Social Education.

RE: Religious Education.

SEN: Special Educational Needs.

SCHOOL PREMISES REGULATIONS (SPR): The Education (School Premises) Regulations 1996, which came into effect on 1 September 1996.

TEACHING AREA: All areas in a school where teaching or learning is accommodated, including timetabled teaching spaces and untimetabled areas such as the library resource centre and other resource areas predominantly for pupil use, darkroom, kiln room, heat treatment bay, group rooms, self-study areas and social areas. It does not include dining or predominantly staff areas.

WORKPLACE: A place to work, i.e: a table space and seat, for one pupil.

Primary

BASIC TEACHING AREA: The class bases and any shared teaching areas, but excluding supplementary teaching areas (see below).

CLASS BASE: A room, bay or open area exclusively used by one class for core activities.

SHARED TEACHING AREA: An area shared by two or more classes for general activities.

SUPPLEMENTARY TEACHING AREA: A room or other discrete space designated for particular activities such as a hall, library, studio or specialist practical areas (whether timetabled or not).

YEARS R TO 6: Primary school years are numbered from R and 1 to 6. Pupils enter the Reception class (Year R) at, or shortly before, age five. Pupils under five may also be in nursery classes or nursery schools.

Secondary

FLA: Foreign Language Assistant.

GNVQ: General National Vocational Qualification: a range of vocational courses usually taken in the sixth form but sometimes at KS4.

PECT: Pneumatics, Electronics and Control Technology: activities in NC design and technology, often done in one type of space.

SELF-STUDY: Private or unsupervised study by pupils, usually post-16, during the timetabled week.

YEARS 7 TO 11: Secondary school years are numbered from 7 (first year) to 11 (end of statutory schooling). The sixth form is sometimes referred to as years 12 and 13.

Grounds

PLAYING FIELDS: an outdoor, grassed area suitable for the playing of team games, laid out for that purpose and capable of sustaining team games for seven hours a week during term time[1].

HARD SURFACED GAMES COURTS: suitable hard surfaced area laid out for court(s) for team games, plus any marginal areas that may be used for small games and skills practice.

INFORMAL AND SOCIAL AREAS: suitable areas for informal games, including infant PE, social interaction and protected leisure activity, of which over half should be hard surfaced.

HABITAT AREAS: planted or specially developed areas to support the curriculum.

BUILDINGS AND ACCESS: total area of the footprints of all school buildings, roads, car parking, service and delivery areas and pedestrian access within the school site.

Note:
1: See DfEE Circular 10/96: The 1996 School Premises Regulations.

Printed in the United Kingdom for HMSO
Dd. 301878 C50 8/96